Energy Efficiency Improvement of Electric Machines without Rare-Earth Magnets

Energy Efficiency Improvement of Electric Machines without Rare-Earth Magnets

Editors

Vladimir Prakht
Mohamed N. Ibrahim
Vadim Kazakbaev

MDPI • Basel • Beijing • Wuhan • Barcelona • Belgrade • Manchester • Tokyo • Cluj • Tianjin

MDPI

Editors
Vladimir Prakht
Ural Federal University
Russia

Mohamed N. Ibrahim
Ghent University
Belgium

Vadim Kazakbaev
Ural Federal University
Russia

Editorial Office
MDPI
St. Alban-Anlage 66
4052 Basel, Switzerland

This is a reprint of articles from the Special Issue published online in the open access journal *Energies* (ISSN 1996-1073) (available at: https://www.mdpi.com/journal/energies/special_issues/Energy-Efficiency-Improvement).

For citation purposes, cite each article independently as indicated on the article page online and as indicated below:

LastName, A.A.; LastName, B.B.; LastName, C.C. Article Title. *Journal Name* **Year**, *Volume Number*, Page Range.

ISBN 978-3-0365-7656-5 (Hbk)
ISBN 978-3-0365-7657-2 (PDF)

Contents

About the Editors

Vladimir Prakht

Vladimir Prakht received a graduate degree in engineering and a Ph.D. degree from the Department of Electrical Engineering, Ural Federal University, Yekaterinburg, Russia, in 2004 and 2007, respectively. He submitted his Ph.D. dissertation on optimal control and mathematical modeling induction heating systems in 2006. He is an Associate Professor at the Department of Electrical Engineering, Ural Federal University. His research interests include the mathematical modeling and optimal design of energy-efficient electric motors and generators.

Mohamed N. Ibrahim

Mohamed N. Ibrahim (Senior member, IEEE) received an M.Sc. degree in electrical power and machines engineering from Tanta University, Egypt in 2012, and a Ph.D. degree in electromechanical engineering from Ghent University, Belgium in 2017. In 2008, he became a Teaching Assistant at the Electrical Engineering Department, Kafrelshiekh University, Egypt and since 2018, he has been an Assistant Professor (on leave). Currently, he is a Post-Doctoral Researcher with the Department of Electromechanical Systems and Metal Engineering, Ghent University, Belgium. He is also an Affiliate Member of Flanders Make, the strategic research center for the manufacturing industry in Flanders, Belgium. His major research interests include the design and control of electrical machines and drives for industrial and sustainable energy applications. Dr. Ibrahim has been a recipient of the Kafrelshiekh University Award several times for his international scientific publications. He serves as a reviewer for several journals and conferences, including *IEEE Trans. on Industrial Electronics*, *Industry Applications*, *Magnetics*, etc. He serves as a Guest Editor for *Energies*, *Mathematics* and *Electronics*.

Vadim Kazakbaev

Vadim Kazakbaev received a graduate degree in engineering and a Ph.D. degree from the Department of Electrical Machines, Ural Federal University, Yekaterinburg, Russia, in 2010 and 2017, respectively. He submitted his Ph.D. dissertation on "Development of High-Performance Synchronous Reluctance Motor" in 2016. He is a Junior Researcher and an Associate Professor with the Department of Electrical Engineering, Ural Federal University. His research interests include electrical engineering, the design of electrical machines, and the control of electrical drives.

Preface to "Energy Efficiency Improvement of Electric Machines without Rare-Earth Magnets"

Electric motors consume about 70% of industrial electricity and about 40%–45% of produced electricity in the world. This means that using high-efficiency electric motors will reduce the level of energy consumption and the environmental impact, resulting in reductions in costs and the emissions of CO_2. In addition, it will significantly reduce the need for new power plants, thus reducing the invested resources to do so. Electric machines with rare-earth magnets have the highest efficiency and power density.

However, rare-earth magnets are expensive, and their manufacturing process, as well as the process of mining rare-earth raw materials, is harmful to the environment. Therefore, the development of energy-efficient electric machines without rare-earth magnets is of great interest. Topics of interest for this Special Issue include, but are not limited to:

- The modeling and design of energy-efficient electric machines without rare-earth magnets;
- Synchronous reluctance machines;
- Permanent-magnet-assisted synchronous reluctance machines;
- Switched reluctance machines;
- Hybrid switched reluctance machines;
- DC-excited flux-switching machines;
- PM-excited flux-switching machines;
- DC-excited synchronous machines;
- Direct-on-line and line start motors;
- Induction machines;
- Electric generators without rare-earth magnets for wind turbines;
- Other electric machines without rare-earth magnets;
- Control techniques for electric machines without rare-earth magnets.

Vladimir Prakht, Mohamed N. Ibrahim, and Vadim Kazakbaev
Editors

energies

Editorial

Energy Efficiency Improvement of Electric Machines without Rare-Earth Magnets

Vladimir Prakht [1], Mohamed N. Ibrahim [2,3,4,*] and Vadim Kazakbaev [1]

1 Department of Electrical Engineering and Electric Technology Systems, Ural Federal University, 620002 Ekaterinburg, Russia; va.prakht@urfu.ru (V.P.); vadim.kazakbaev@urfu.ru (V.K.)
2 Department of Electromechanical, Systems and Metal Engineering, Ghent University, 9000 Ghent, Belgium
3 Flanders Make @UGent—Core Lab MIRO, 3001 Leuven, Belgium
4 Electrical Engineering Department, Kafrelshiekh University, Kafrelshiekh 33511, Egypt
* Correspondence: mohamed.ibrahim@ugent.be

Electric motors are one of the largest consumers of electricity and are responsible for 40–45% of the world's energy consumption. In some applications, this proportion is even higher. For example, in industry, electric motors are responsible for about 70% of electricity consumption. Therefore, reducing the energy consumption of electric motors is critical both in terms of saving limited fossil fuels and in terms of reducing the load on national electricity networks. In addition, reducing energy consumption is important in terms of mitigating the impact of human activities on the environment. It is known that the magnetic flux generated by a coil with current is proportional to the product of the area of the pole covered by the coil and the cross-section of the winding wire, that is, the linear dimensions of the coil to the fourth power. The flux created by a permanent magnet is proportional to its volume, that is, its linear dimensions to the third degree. For this reason, electrical machines with high-coercivity rare-earth permanent magnets are potentially the most compact and energy efficient. At the same time, the cost of rare earth magnets is high, they are supplied by only a few suppliers, the raw material extraction process for their manufacture is harmful to the environment and there are problems with disposal. This gives good motivation for developing electric machines without rare earth magnets [1–10].

This Special Issue aimed to gather new research publications on various topics related to improving the energy efficiency of electric machines without using rare-earth magnets. Ten articles have been published that cover various topics. In [1], it is shown that synchronous motors without permanent magnets can be an energy efficient alternative to permanent magnet motors, as well as to conventional induction motors in widespread applications such as fixed-speed electrically driven pump units. The energy-efficiency indicators of a direct-on-line synchronous reluctance motor with a power rating of 4 kW are compared with a motor with rare-earth permanent magnets on the rotor, as well as with an IE3 energy efficiency class induction motor. For this analysis, the performance curves of the motors are interpolated based on experimental data. To consider the influence of the power factor of the motors with direct power supply from the grid, losses in the supply cable of the pumping station were also taken into account. It is shown that despite the increased losses in the supply cable due to a reduced power factor, the synchronous reluctance motor provides a significant energy-saving effect compared to the induction motor due to much higher motor efficiency. The permanent magnet motor provides slightly greater energy savings; however, unlike the synchronous reluctance motor, it has a long payback period due to its high cost associated with the use of expensive rare earth magnets. The study in [2] is devoted to increasing the energy efficiency of a solar water pump with an electric drive by optimizing the parameters of its PI controllers. The solar pump is a self-powering system that does not require power from the utility network. The considered electric drive uses an induction motor that does not have expensive permanent magnets.

Citation: Prakht, V.; Ibrahim, M.N.; Kazakbaev, V. Energy Efficiency Improvement of Electric Machines without Rare-Earth Magnets. *Energies* 2023, *16*, 3573. https://doi.org/10.3390/en16083573

Received: 2 March 2023
Revised: 6 April 2023
Accepted: 18 April 2023
Published: 20 April 2023

The control system uses DC link voltage and motor speed PI controllers. It is shown that, when tuning the PI controllers, the coyote optimization algorithm is more efficient in terms of computational costs and the final result than the widely used Ziegler–Nichols and trial error tuning algorithms. Paper [3] theoretically discusses the design of an axial flux induction motor without permanent magnets, the magnetic core of which is made of Soft Magnetic Composites (SMC), which simplifies its production. It is shown that such a motor can correspond to the IE3 energy efficiency class. It is also highlighted that rotor skew can provide low torque ripple for such a motor. In [4], the choice of the best flux reversal machine (FRM) configuration for a small wind generator with inexpensive ferrite magnets is discussed. A three-phase FRM design with six stator winding poles and eight rotor teeth with significantly reduced cogging torque and torque ripple is selected for this application based on comparative finite element analysis (FEA). In [5], manual optimization of the parameters of the rotor geometry of a synchronous reluctance motor without magnets with a direct start from the grid is carried out. As a result, a significant increase in efficiency and power factor is achieved. The results of testing the motor at the rated load are provided, verifying the theoretical results obtained. In [6], two winding layouts of a three-phase switched reluctance motor (SRM) with twelve stator teeth and eight rotor teeth are compared in terms of power loss and efficiency. It is shown that the winding with a short pitch provides a lower copper of loss than the winding with a full pitch. Comprehensive comparative experiments were carried out to verify the results obtained using FEA. In [7], various aspects of the development and manufacture of an electric motorcycle electric drive based on an SRM are discussed. The choice of SRM configuration for this application is discussed. It is shown that a three-phase SRM with six stator teeth and four rotor teeth shows the best performance for this application. The dynamic simulation of SRM as part of the drive of the electric vehicle is carried out. A method for determining the position of the rotor and its experimental implementation are discussed. The test results of the SRM under consideration on the test bench and as part of an electric motorcycle are also provided. In [8], a novel design of a linear flux switching motor (FSM) is proposed employing inexpensive ferrite magnets with a modular passive stator for railway transport. Optimization of the FSM parameters using a genetic algorithm was carried out. The results of the electromagnetic and thermal analysis are presented. The proposed design of the linear FSM has a high specific thrust and high efficiency and a wide range of velocity control. In [9], a novel FSM design without permanent magnets with an external rotor is proposed. The outer rotor is passive and has no windings. Due to the selected shape of the stator teeth, the proposed FSM has a reduced leakage flux, high power factor and low torque ripple. The FSM design is optimized using a genetic algorithm. The results of electromagnetic and thermal finite element analysis are presented. A theoretical comparison of the performance of the optimized FSM with several conventional wound rotor synchronous machines and FSM configurations is provided, which shows the superiority of the proposed solution in terms of efficiency and specific torque. In [10], for electric vehicles, a novel design of an SRM with twelve stator teeth and eight rotor teeth without permanent magnets is proposed. Unlike conventional SRMs, the proposed motor uses a hybrid winding excitation algorithm that provides increased specific torque. The results of FEA are confirmed by comprehensive experimental verification carried out at various speeds and torques of the motor. It is shown that the specific torque and efficiency of the proposed hybrid SRM are significantly higher than that of the conventional one.

Author Contributions: Conceptualization, M.N.I., V.P. and V.K.; methodology, M.N.I.; software, V.K.; validation, V.K., M.N.I. and V.P.; formal analysis, V.K.; investigation, V.K.; resources, V.P.; data curation, V.K.; writing—original draft preparation, M.N.I.; writing—review and editing, M.N.I., V.K. and V.P.; visualization, V.K.; supervision, V.P. All authors have read and agreed to the published version of the manuscript.

Acknowledgments: The editors thank the authors for submitting their studies to the present Special Issue which led to a successful completion. We deeply acknowledge the Energies Reviewers for enhancing the quality and impact of all submitted papers. Finally, we sincerely thank the editorial staff of Energies for their stunning support during the development and publication of this Special Issue.

Conflicts of Interest: The author declares no conflict of interest.

References

1. Kazakbaev, V.; Prakht, V.; Dmitrievskii, V.; Oshurbekov, S.; Golovanov, D. Life Cycle Energy Cost Assessment for Pump Units with Various Types of Line-Start Operating Motors Including Cable Losses. *Energies* **2020**, *13*, 3546. [CrossRef]
2. Arfaoui, J.; Rezk, H.; Al-Dhaifallah, M.; Ibrahim, M.N.; Abdelkader, M. Simulation-Based Coyote Optimization Algorithm to Determine Gains of PI Controller for Enhancing the Performance of Solar PV Water-Pumping System. *Energies* **2020**, *13*, 4473. [CrossRef]
3. Arabul, F.K.; Senol, I.; Oner, Y. Performance Analysis of Axial-Flux Induction Motor with Skewed Rotor. *Energies* **2020**, *13*, 4991. [CrossRef]
4. Manne, B.; Kumar, M.K.; Akuru, U.B. Design and Performance Assessment of a Small-Scale Ferrite-PM Flux Reversal Wind Generator. *Energies* **2020**, *13*, 5565. [CrossRef]
5. Kim, H.; Park, Y.; Oh, S.T.; Jang, H.; Won, S.H.; Chun, Y.D.; Lee, J. A Study on the Rotor Design of Line Start Synchronous Reluctance Motor for IE4 Efficiency and Improving Power Factor. *Energies* **2020**, *13*, 5774. [CrossRef]
6. Pupadubsin, R.; Kachapornkul, S.; Jitkreeyarn, P.; Somsiri, P.; Tungpimolrut, K. Investigation of Torque Performance and Flux Reversal Reduction of a Three-Phase 12/8 Switched Reluctance Motor Based on Winding Arrangement. *Energies* **2022**, *15*, 284. [CrossRef]
7. Kachapornkul, S.; Pupadubsin, R.; Somsiri, P.; Jitkreeyarn, P.; Tungpimolrut, K. Performance Improvement of a Switched Reluctance Motor and Drive System Designed for an Electric Motorcycle. *Energies* **2022**, *15*, 694. [CrossRef]
8. Jan, H.U.; Khan, F.; Ullah, B.; Qasim, M.; Milyani, A.H.; Azhari, A.A. Design and Thermal Analysis of Linear Hybrid Excited Flux Switching Machine Using Ferrite Magnets. *Energies* **2022**, *15*, 5275. [CrossRef]
9. Akbar, S.; Khan, F.; Ullah, W.; Ullah, B.; Milyani, A.H.; Azhari, A.A. Performance Analysis and Optimization of a Novel Outer Rotor Field-Excited Flux-Switching Machine with Combined Semi-Closed and Open Slots Stator. *Energies* **2022**, *15*, 7531. [CrossRef]
10. Abhijith, V.; Hossain, M.J.; Lei, G.; Sreelekha, P.A.; Monichan, T.P.; Rao, S.V. Hybrid Switched Reluctance Motors for Electric Vehicle Applications with High Torque Capability without Permanent Magnet. *Energies* **2022**, *15*, 7931. [CrossRef]

energies

MDPI

Article

Life Cycle Energy Cost Assessment for Pump Units with Various Types of Line-Start Operating Motors Including Cable Losses

Vadim Kazakbaev [1], Vladimir Prakht [1,*], Vladimir Dmitrievskii [1], Safarbek Oshurbekov [1] and Dmitry Golovanov [2]

1 Department of Electrical Engineering and Electric Technology Systems, Ural Federal University, Ekaterinburg 620002, Russia; vadim.kazakbaev@urfu.ru (V.K.); vladimir.dmitrievsky@urfu.ru (V.D.); safarbek.oshurbekov@urfu.ru (S.O.)

2 Department of Electrical Engineering, University of Nottingham, Nottingham NG7 2RD, UK; dmitry.golovanov@nottingham.ac.uk

* Correspondence: va.prakht@urfu.ru

Received: 11 June 2020; Accepted: 6 July 2020; Published: 9 July 2020

Abstract: The paper presents a comparative analysis of life-cycle energy consumption for three different types of 4 kW line-start motors used in a pump unit with throttling: the most widely used induction motor with IE3 efficiency class, line start permanent magnet synchronous motor with IE4 efficiency class and line start synchronous reluctance motor with IE4 efficiency class. The operating cycle for pump units with constant flow is considered for the above-mentioned types of motors taking into account not only the losses in the pump and motor, but also in the power supply cable. It is shown that the line start synchronous reluctance motor without magnets has the highest efficiency over the entire considered loading range. However, its power factor is lower than that of the synchronous motor with magnets and therefore it has more significant losses in power supply cable. Despite this disadvantage, the line-start reluctance motor is a good alternative to widespread induction motor since it allows saving of approximately 4000 euro more than the latter during the 20 years life cycle. It also provides similar savings in comparison to the permanent magnet synchronous motor, but unlike it, it does not have costly rare-earth materials in the rotor.

Keywords: centrifugal pump; energy efficiency; induction motor; line-start synchronous motor; synchronous reluctance motor; throttling control

1. Introduction

Nowadays, most of the drives with the direct start from the mains power supply utilize induction motors (IM) with aluminium squirrel cage. Typically, IM with casting squirrel cage has a relatively low class of efficiency IE3 [1,2].

Although it is currently acceptable to use the motors with the efficiency class IE3 for industrial applications, according to the EU plans, the efficiency requirements will be increased in the future. For instance, starting from 1st of July 2023, the motors in the EU with the output power greater than 75 kW should comply with the IE4 class of efficiency [3]. Future plans imply the expansion of IE4 requirements to the motors with lower output power (<75 kW) and moving towards the IE5 class of efficiency for high output power motors [4]. In addition, even at the present time, the use of motors of IE4 and IE5 classes can be feasible, due to constantly increasing cost of energy resources and the need to reduce environmental impact [5].

Therefore, further efficiency improvement of IM is necessary in order to comply with upcoming changes in energy efficiency standards. One of the possible ways is to replace the aluminium squirrel cage with a copper one. However, the latter is more expensive in comparison to the aluminium squirrel cage [2]. For that reason, the IMs with copper squirrel cage are not so widespread for the drives with direct start from the mains power supply.

Recently, the major industrial companies [6,7] have developed synchronous permanent magnet motors with the line start from the mains power supply—so-called line start permanent magnet synchronous motors (LS-PMSM), which comply with the IE4 class efficiency. Such synchronous motors have a squirrel cage type winding on the rotor. This squirrel cage allows starting the motor directly from the mains power supply in asynchronous mode. The rotor accelerates until its frequency gets close to the synchronous frequency. In synchronous mode, the squirrel cage dampens rotor vibrations during sudden changes in load.

There is a number of publications dedicated to the development of LS-PMSM [8–13]. However, LS-PMSM cannot compete with IM contribution to industrial use due to the high cost of the former. Another problem in the manufacturing of LS-PMSM is its technological dependence on rare earth materials suppliers from China. About 95% of the total amount of rare earth raw materials worldwide comes from China [14]. Since this situation is unlikely to change in the foreseeable future, there is always a threat of an unstable change in the price of raw materials for the production of permanent magnets. The already high prices for materials for the production of rare earth magnets can change several times within a few years [14,15].

In addition, the rare earth mineral processing technology is associated with significant environmental damage. It is claimed that the processing of each ton of rare-earth element concentrate results in generation of 1–1.4 t of radioactive waste. Only a small part of the resulting slush contains rare-earth elements and is subsequently extracted for refinement [16].

In view of the above-noted disadvantages of the motors with rare earth magnets, an alternative to the LS-PMSM must be sought. Currently, synchronous reluctance motors (SynRM), which have no magnets and no short-circuited startup winding in the rotor, and are powered by a variable frequency drive (VFD) are the good alternatives to IM [17–21]. They meet the criteria of IE4 energy efficiency [22,23] and IE5 [19,24,25], according to the standard IEC 60034-30-2 «Rotating electrical machines – Part 30-2: Efficiency classes of variable speed AC motors (IE-code)». These kinds of SynRMs were recently launched for production by large industrial companies. [22,26,27].

There are also IE4 energy efficiency class synchronous reluctance motors that are powered directly from the mains power supply, the rotor of which is manufactured with a short-circuited startup winding (line start synchronous reluctance motor, LS-SynRM) [28–31]. These kinds of LS-SynRMs have a higher energy efficiency class than IM as well as approximately the same manufacturing cost [32].

Wide use of energy-efficient LS-SynRMs can lead to a decrease in energy consumption and energy intensity of GDP, as well as lower Greenhouse Gas emissions during electricity generation. The use of energy-efficient LS-SynRM instead of IM will help to achieve the goals stated in the energy and environmental strategies of the European Union (European Green Deal [5]), USA (State Energy Program [33]), Switzerland (supporting Paris Climate Agreement [34]), China (supporting Paris Climate Agreement [35]), Japan (Net Zero Energy Building [36]), South Korea (supporting Paris Climate Agreement [37]) and other countries.

Pump systems consume almost 22% of all electric energy generated throughout the world [38]. Most of the pump drives are powered directly from the mains power supply [39,40]. This suggests the high energy-saving potential of LS-SynRM in pump applications.

A large number of works are dedicated to the comparison of the energy consumption of pumping systems with various types of motors (induction motors, synchronous motors with rare earth permanent magnets, synchronous reluctance motors without magnets). However, all these works are dedicated to pumping systems with motor frequency control using variable frequency drive (VFD) [17,21,41–43].

Energy-saving effect of using various types of motors in pumping systems powered directly from the mains power supply are considered much less frequently. Thus, in [40] the comparison of the energy consumption for LS-PMSM and IM with the direct feed from the mains power supply of classes

IE3 and IE4 in a centrifugal pump with a throttling control is considered. In that work, it was shown that when choosing the motor, it is necessary to take into account not only the energy efficiency class, but also the efficiency under reduced loads.

However, the estimation of the energy-saving effect of LS-SynRM use in pumping systems, and their comparison to IM and LS-PMSM, including the energy losses in power supply cable, has not yet been considered in the literature to the best of our knowledge. The energy losses in the power supply cable depend on the power factor. The power factor should be taken into account due to the increased reactive component of the current, which contributes significantly to the increase of the total current. In [44–46], it is noted that LS-SynRM has a small power factor, which can lead to losses in the power supply cable. Since the motors studied in this article have different values of the power factor, not only their efficiency but also their influence of the power factor on losses in the power supply cable is taken into account when comparing their energy consumption.

In this paper, a comparative assessment of the energy consumption for a low-power pump drive (4 kW) with various types of electric motors is carried out. The following electric motor types are considered: LS-SynRM with energy efficiency class IE4 (Figure 1a), LS-PMSM with energy efficiency class IE4 (Figure 1b) and IM with energy efficiency class IE3 (Figure 1c). All three considered motors have a similar design of the stator. However, their rotor design is different. All the motors have a starter cage on the rotor for the asynchronous startup. However, LS-SynRM has also the magnetic anisotropy of the rotor structure formed by magnetic barriers. Therefore, it enters into synchronous operation after starting. LS-PMSM also goes into synchronous operation due to the synchronizing torque produced by the permanent magnets installed in the rotor. The synchronous motors (LS-SynRM and LS-PMSM) usually have a better efficiency compared to IM due to the reduced rotor losses. Various loading modes of the duty cycle of the pump drive with unregulated rotation speed are taken into account. The costs of electric energy consumption during the pump life-cycle is chosen as the main criterion for comparing the motors.

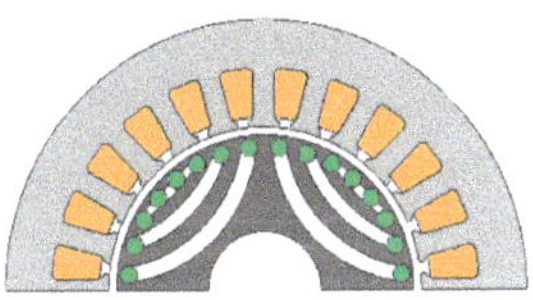
(a)

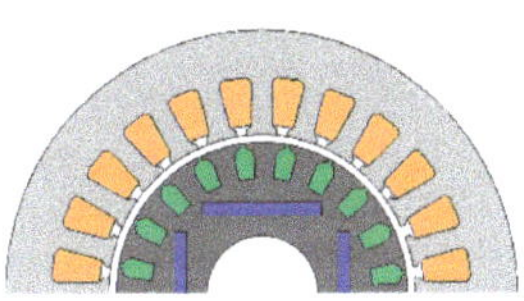
(b)

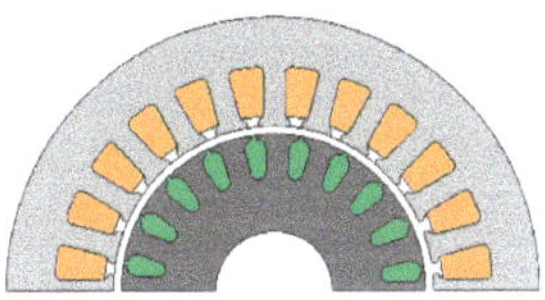
(c)

Figure 1. Schematic representation of motor geometry: (**a**) line start synchronous reluctance motors (LS-SynRM); (**b**) line start permanent magnet synchronous motors (LS-PMSM); (**c**) induction motors (IM).

It must be highlighted that the analysis presented in this paper is not based on theoretical calculations of motor performances using a modelling software. Instead, an approach to the analysis of energy consumption based on data from manufacturers' datasheets and experimental data is used in the work. For IM and LS-PMSM, the data from manufacturers datasheets are used [47,48]. Since LS-SynRMs of high energy efficiency classes are not yet mass-produced, the experimental data for LS-SynRM are taken from the article [28]. More specifically, the data on the efficiency and power factor of the motors are used in the present analysis. For data processing, polynomial interpolation is applied to the entire load range under consideration. Interpolation of experimental data and other calculations were performed in the MATLAB program and are described in Formulas (1)–(9). The proposed calculation method can be used for choosing the type of electric motor in pumping systems.

2. Evaluating Pump Energy Consumption

The mathematical model of pump unit for calculating the energy consumption of a pump drive with various types of the motors is presented below. The schematic representation of the drive of a pump unit with one electric motor, powered directly from the mains power supply, is shown in

Figure 2. The electric motor is fed directly from the mains and is coupled to a centrifugal pump without intermediate mechanical gears. The required motor electrical power $P_{motor\ el}$ depends on the flow Q [39]:

$$P_{motor\ el} = \eta_{motor} \cdot P_{mech}; \tag{1}$$

$$P_{mech} = \eta_{pump} \cdot P_{hydr} = \eta_{pump} \cdot \rho \cdot g \cdot H \cdot Q = f(Q) \tag{2}$$

where H—water pressure, is defined from the H-Q characteristic of the pump from the catalogue; g—acceleration of gravity; ρ—density of a liquid; P_{hydr}—hydraulic power of pump; P_{mech}—mechanical power of pump, defined from characteristic $P_{mech} = f\ (Q)$ from catalogue; η_{pump}—pump efficiency; η_{motor}—motor efficiency.

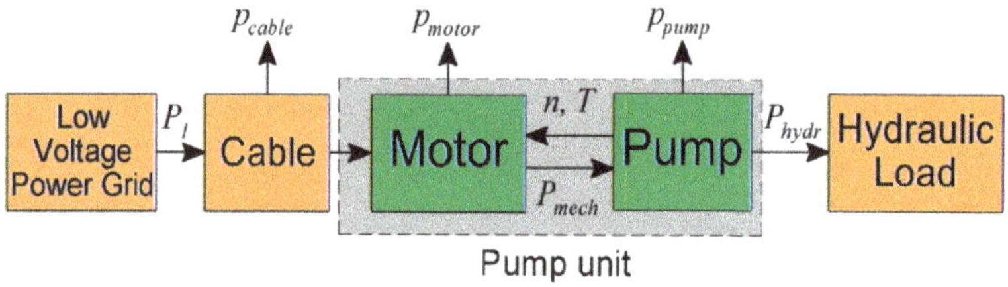

Figure 2. Diagram of a single pump unit for fixed-speed operation. In this diagram: n, T are the rotational speed and load torque of the motor; p_{motor} is the motor loss; p_{pump} is the pump loss.

In addition to losses in the motor, losses in the power supply cable are also taken into account. The power supply cable electrical loss p_{cable} depends on the active phase resistance of the cable and the value of the motor current [49]:

$$p_{cable} = 3 \cdot R_{cable} \cdot I_{motor}^2, \tag{3}$$

where R_{cable}—cable phase resistance; I_{motor}—motor current.

The motor current is calculated according to formula:

$$I_{motor} = P_{mech} / (\sqrt{3} \cdot V_{motor} \cdot \cos\varphi \cdot \eta_{motor}), \tag{4}$$

where V_{motor} = 400 V; $\cos\varphi$ and η_{motor}—power factor and motor efficiency, according to data from the catalogue.

Considering losses in cable, the electric power P_1 consumed from the mains by the pump unit is calculated as:

$$P_1 = \eta_{motor} \cdot P_{mech} + p_{cable}. \tag{5}$$

To calculate the energy consumption of a pump drive, the pump performance characteristics from the manufacturer's datasheets were used [50]. In order to compare the energy consumption of the electric motors as a part of the pump unit when controlling flow by a throttling valve, a centrifugal pump B-NM4 65/25B/B (manufactured by Calpeda S.p.A., Montorso Vicentino, Vicenza, Italy) with the rated power P_{rate} = 4 kW and with rated rotational speed n = 1450 rpm was used [50]. Pump data are shown in Table 1. Q_{BEP} denotes the flow at the best efficient point (BEP), and H_{BEP} denotes the pump head at BEP.

Table 1. Published characteristics of the pump from manufacturer.

Type	Rated Mechanical Power, W	Rate Rotational Speed, rpm	Q_{BEP}, m^3/h	H_{BEP}, m	Pump Efficiency in BEP, %
B-NM4 65/25B/B	4000	1450	60	15.4	75.5

The calculation was performed for three different four-pole electric motors with the rated power of 4 kW, namely LS-SynRM (a test prototype [28]), LS-PMSM (manufacturer WEG [47]) and IE3 efficiency class induction motor (IM, manufacturer WEG [48]). For the serially produced LS-PMSM and IM, datasheets on their efficiency are used.

There are still no commercially available high-performance LS-SynRMs on the market to the best of our knowledge. ABB Group corporation has announced the launch of IE4 class LS-SynRM [51], however, at the moment of writing this manuscript, these motors are still not available on the market. Therefore, to perform the calculation of LS-SynRM power consumption, the data of the experimental sample described in [28] were used.

Efficiency data for the motors are shown in Tables 2 and 3 and in Figure 3. Current data of the motors are shown in Figure 4. Power factor data of the motors are shown in Figure 5. The motor current can be found based on the efficiency and power factor data using Equation (4).

Table 2. Motor characteristics.

Type of Motor	Rated Mechanical Power, W	Poles	Frame Size	Frame Material	Weight, kg	Rated Voltage, V
LS-SynRM	4000	4	IEC 112	No data	No data	400
LS-PMSM	4000	4	IEC 112	Cast iron	49	400
IM	4000	4	IEC 112	Cast iron	42	400

Table 3. Motor characteristics.

Type of Motor	Motor Efficiency, %			Motor Power Factor		
	50% Load	75% Load	100% Load	50% Load	75% Load	100% Load
LS-SynRM	91.6	92.4	91.9	0.607	0.713	0.755
LS-PMSM	89.0	91.0	91.7	0.68	0.81	0.88
IM	88.7	89.1	88.8	0.6	0.72	0.78

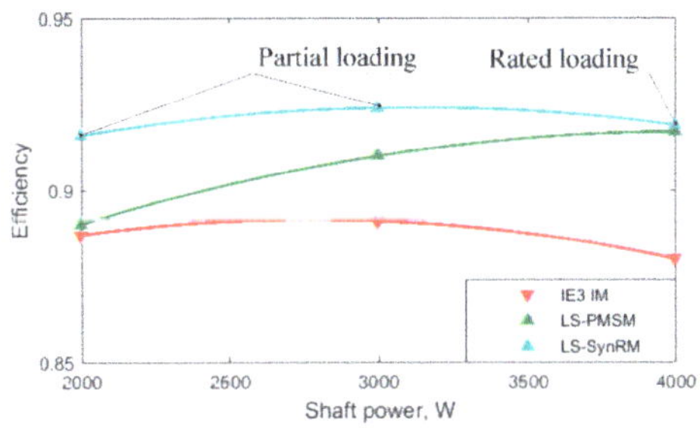

Figure 3. Motor efficiency.

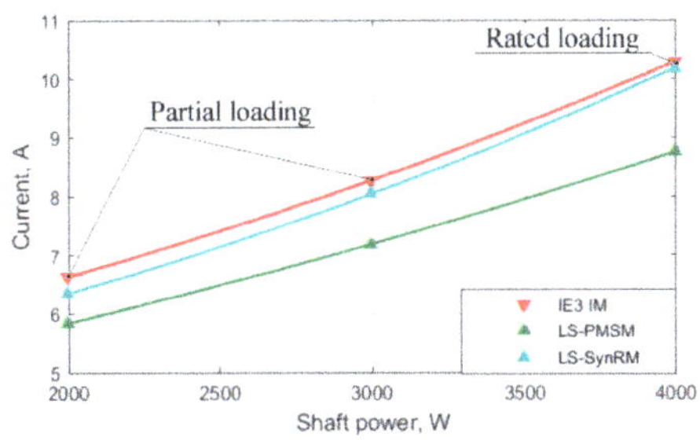

Figure 4. Motor current.

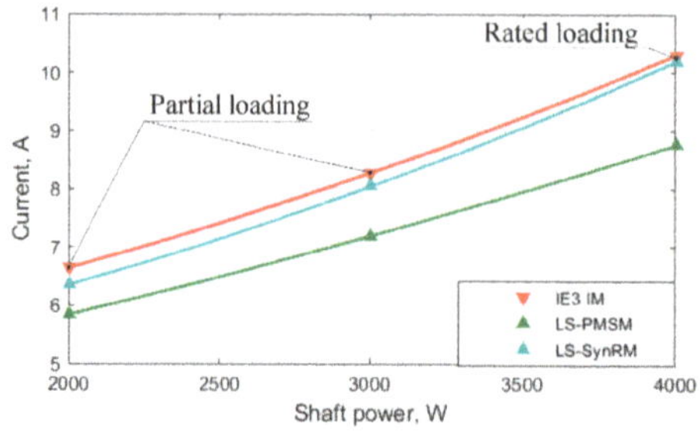

Figure 5. Motor power factor.

Note that all three considered motors have approximately the same size and are located in standard casings IEC 112 with mounting dimensions in accordance with IEC standard 60072-1-1991. All data for LS-PMSM and IM are taken from the datasheets of the manufacturer [47,48]. Data for LS-SynRM are taken from the article [28].

In Figures 3–5, the data on motor efficiency, current and power factor are presented. It can be seen that the LS-SynRM has the highest efficiency values in various modes (Figure 3). The rated efficiency of the LS-SynRM is 91.9%. Also, LS-PMSM is more efficient than the traditional IM. According to the data in Table 3, the estimated efficiency of the IM is only 88.8%, while the efficiency of LS-PMSM is significantly higher at 91.7%. Figure 3 also demonstrates that the LS-PMSM efficiency is higher than that of the IM over the entire considered loading range.

The efficiency of the LS-PMSM in the nominal mode is the same as that of the LS-SynRM. However, in partial load modes, the LS-PMSM's efficiency decreases much faster than the one of LS-SynRM. LS-PMSM has the highest power factor. The power factors of IM and LS-SynRM have approximately the same values (0.78 and 0.755, respectively) which are significantly lower than the one for PMSM (0.88). The rate of the power factor decrease with a decrease in load is approximately the same for all studied motors: when the load was decreased from 100% to 50%, the power factor decreased by approximately 0.15.

3. Pump Operating Cycle

The operation of the pump unit was considered in the loading points wherein the water flow over the duty cycle changes in accordance with the typical characteristic of fixed-speed pump applications, as shown in reference [52]. In reality, however, even in the case of fixed-speed pumps, the flowrate is very seldom constant. For example, even in a simple pump system when a pump is used to move liquid from one reservoir to another, the duty points vary due to the level of the reservoirs, which means the pump does not operate all the time at the best efficiency. A typical duty cycle of a pump with an approximately constant flowrate described in [52] is characterized by three discrete modes (Figure 6).

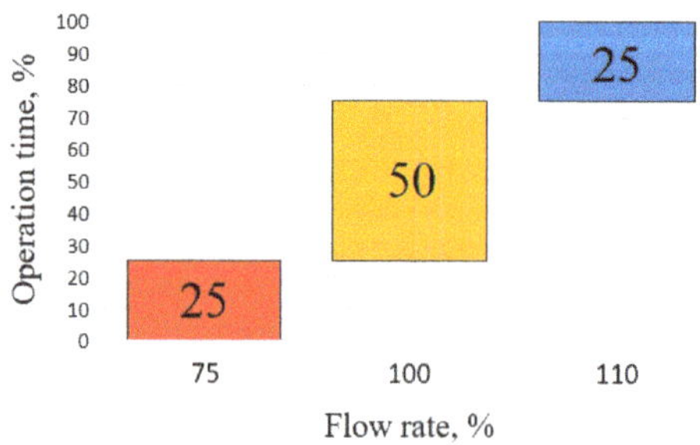

Figure 6. Flow-time profile. The numbers in the rectangles indicate the share of the operation time of a loading point.

An electric motor is connected directly to the mains power supply. Therefore, the flow Q of the pump is adjusted using a throttle. In this case, the water pressure changes in accordance with the Q-H curve of the pump, and the operating point is the intersection point of the pump characteristic (red line in Figure 7a) and the hydraulic system characteristic (blue, green and emerald lines in Figure 7a). Figure 7a shows the interpolation results of the Q-H characteristic of the selected pump according to manufacturer data [50].

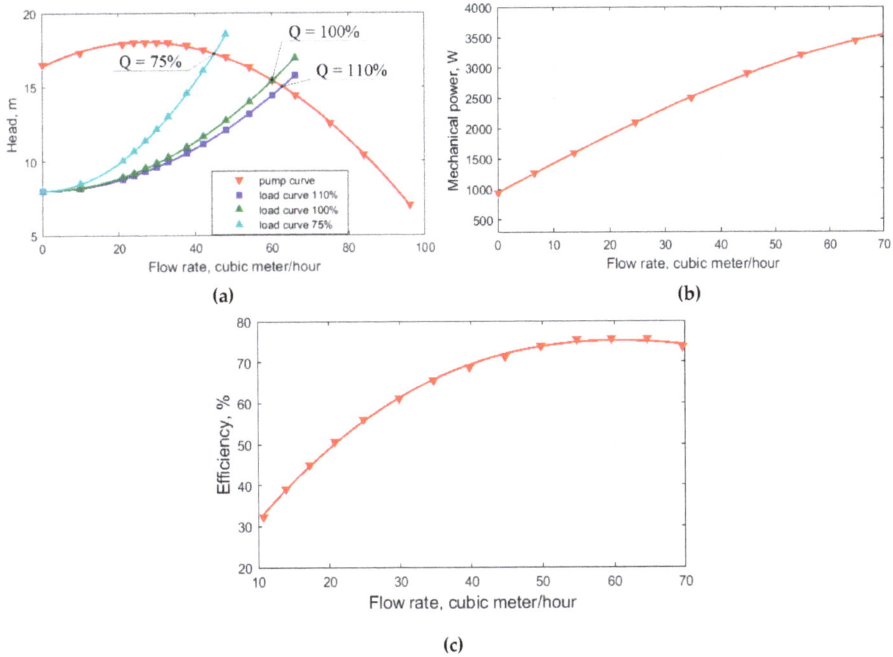

Figure 7. Pump performances: (**a**) Q-H curve; (**b**) pump mechanical power; (**c**) pump efficiency.

The mechanical (input) power curve of the pump as a function of the flow is reported by the pump manufacturer (Figure 7b). The pump power was determined from the curve in three operation modes (75%, 100% and 110% of the pump flow). The flow corresponding to 100% of the maximum flow in the pump operating cycle was determined based on the pump efficiency curve (Figure 7c) [50], corresponding to maximum efficiency (best efficiency point: $Q = 60\text{m}^3/\text{h}$, $\eta_{pump} = 0.755$).

The efficiencies of the electric motors (Table 2) at the four operation modes of the pump unit were determined using polynomial interpolation of their specification data.

The obtained efficiency values are provided in Table 4, which also shows the following values for each operating mode: the flow, the pump head, and the mechanical output power of the electric motors as percentages of the rated output.

Table 4. Characteristics of pump duty cycle.

Q, %	Q, m^3/h	H, m	P_{mech}, W	T, N·m	η_{pump}, %	η_{motor}, %		
						LS-SynRM	**LS-PMSM**	**IM**
110	66	14.4	3453	22.0	75	0.924	0.915	0.89
100	60	15.4	3335	21.2	75.5	0.925	0.914	0.891
75	45	17.25	2962	18.9	71.4	0.925	0.909	0.891

4. Cable Losses Depending on Motor Power Factor

An important energy parameter for a motor with a feed from the mains power supply is not only the efficiency, but also the power factor, because the reactive current fed to a motor flows not only through its winding, but also through a network of elements from which a motor receives power [53] causing additional losses. Since the considered motors have different power factors and total currents (see Figure 4 and Table 5), it is also necessary to evaluate the influence of this factor on the cost of electricity for the consumer.

Table 5. Motor current at three points of the duty cycle for various pump modes.

Q, %	P_{mech}, W	I_{motor}, A		
		LS-SynRM	**LS-PMSM**	**IM**
110	3453	7.33	6.42	7.42
100	3335	7.12	6.27	7.23
75	2962	6.51	5.84	6.69

Let us have a look at the simplest case when a pump is connected directly to a three-phase 400 V network. In this case, if we want to consider cable losses, we need to take into account the magnitude of the motor current too. For industry, the typical cable length for connecting low-voltage power equipment is about 100 m [54]. In low-power electrical systems with a current load of up to 15A, cables with a cross section of 1.5 mm^2 are typically used [55]. The specific resistance of one phase of a copper cable with these parameters is approximately ρ_{cable} = 12.6 Ohm/km. When performing calculations for the stranded cables of a small cross section, the reactance is usually neglected. We assume that the motor is powered via a typical cable of 100 m length (l_{cable} = 100 m). The phase resistance of such a cable will be $R_{cable} = l_{cable} \cdot \rho_{\text{cable}} = 0.1 \cdot 12.6 = 1.26$ Ohm. Losses in the cable are calculated from (3). The results of interpolation of motor current values for various studied pump modes are presented in Table 5.

Table 6 presents the results of the calculation of losses in the cable according to Formula (3), as well as the values of the losses in the motor, according to the data in Table 4.

Table 6. Losses in motor and cable at three points of the duty cycle for various pump modes.

Q, %	P_{mech}, W	p_{motor}, W			p_{cable}, W		
		LS-SynRM	**LS-PMSM**	**IM**	**LS-SynRM**	**LS-PMSM**	**IM**
110	3453	283.6	320.8	426.8	203.0	155.6	208.0
100	3335	272.4	313.8	408.0	191.8	148.6	197.7
75	2962	241.9	296.6	362.4	160.0	128.8	169.4

In accordance with the assumptions made, the estimated losses in cable are comparable with the ones in the motor (Table 6). At the same time, the cable losses for LS-PMSM are lower by approximately 25% than for LS-SynRM and IM. The main reason of the reduced cable losses for LS-PMSM is its reduced total current due to its higher power factor, according to Equations (3) and (4). These results confirm the importance of the power factor increase for reduction of the energy consumption of the line-start motors.

5. Electric Energy Consumption Results and Discussion

Using the results obtained in the previous sections, we compared the energy consumption of the pump drive with various motors: LS-SynRM, LS-PMSM and IM motor. The daily energy consumption for each electric motor over a full duty cycle of the pump unit in accordance with the corresponding load profile (Figure 6) is determined as:

$$E_{day} = t_{\Sigma} \cdot \sum_{i=1}^{3} (P_{1i} \cdot t_i / t_{\Sigma}). \tag{6}$$

where $i = 1 \ldots 3$ is the number of a loading point; $P_{1\,i}$ is the eclectic power P_1 in a loading point; t_i/t_{Σ} is the share of the operation time of a loading point.

For the year-round operation of the pump unit, the annual energy consumption was calculated as follows:

$$E_{year} = E_d \cdot 365. \tag{7}$$

The cost of electricity consumed (in Euro), considering the applied grid tariffs GT = 0.2036 €/kW·h for non-household consumers [56] for Germany in the second half of 2019, was calculated as follows:

$$C_{year} = E_y \cdot GT. \tag{8}$$

The whole life cycle of pump units is usually about 20 years [57,58]. The energy cost for the lifespan of n = 20 years is assessed without taking into account the maintenance costs and the initial cost of the motors since the market cost of the motors depends on many factors and this was beyond of the topic of the present paper. Furthermore, the pump lifetime expenses often consist mostly of the energy cost (>50–60%) [57,58]. The net present value (NPV) of the life-cycle cost was calculated as follows:

$$C_{LCC} = \sum_{i=1}^{n} \left(C_{year} / [1 + (y - p)]^i \right), \tag{9}$$

where y— interest rate (y = 0.04); p—expected annual inflation (p = 0.02); n—lifetime of the pump unit (n = 20 years) [58].

The results of calculations based on (1)–(9) excluding and including cable losses are presented in Tables 7 and 8, respectively. Figure 8 presents calculated values of life cycle energy cost C_{LCC} for both cases. The energy savings of LS-SynRM and LS-PMSM is assessed in relation to the induction motor.

Table 7. Calculation of energy consumption excluding cable losses.

	t_i/t_{Σ}, %	LS-SynRM	LS-PMSM	IM
P_1, W	25	3736.7	3773.9	3879.9
	50	3607.3	3648.8	3742.9
	25	3204.5	3259.2	3325.0
E_{day}, kW·hour		84.9	86.0	88.1
E_{year}, kW·hour		31,001	31,384	32,173
Annual energy savings, kW·hour		1171	789	–
Annual energy savings, %		3.65	2.45	–
Cost savings, € (per year)		238.5	160.6	–
Cost savings, € (per 20 years)		3900	2626	–
Life cycle energy cost C_{LCC}, € (20 years)		103,208	104,482	107,108

Table 8. Calculation of energy consumption including cable losses.

	t_i/t_Σ, %	LS-SynRM	LS-PMSM	IM
P_1, W	25	3939.7	3929.5	4087.9
	50	3799.1	3797.3	3940.7
	25	3364.5	3387.9	3494.4
E_{day}, kW·hour		89.4	89.4	92.8
E_{year}, kW·hour		32,636	32,658	33,865
Annual energy savings, kW·hour		1228.9	1207.6	–
Annual energy savings, %		3.6	3.6	–
Cost savings, € (per year)		250.2	245.9	–
Cost savings, € (per 20 years)		4181	4020	–
Life cycle energy cost C_{LCC}, € (20 years)		108,651	108,722	112,742

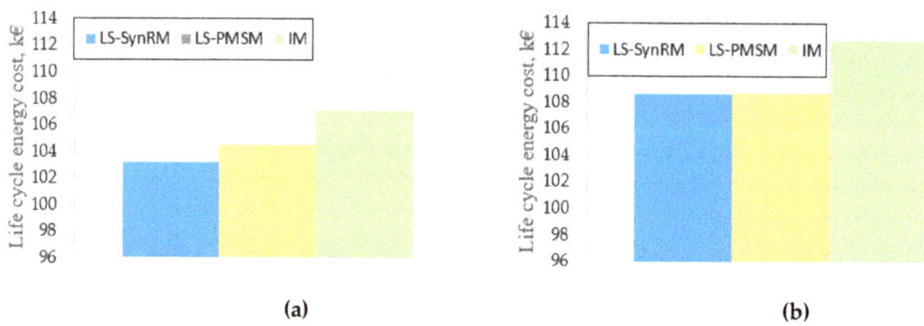

Figure 8. Life cycle energy cost (20 years) using various motor types: (**a**) without considering cable loss; (**b**) considering cable loss.

From Tables 7 and 8 and Figure 8 we can see that without taking into account the cable losses, LS-SynRM provides the lowest energy consumption (E_{day} = 84.9 and E_{year} = 31,001 kW·hour) which is 3.65% more annual energy saving than for IM. For the LS-PMSM and IM the daily energy consumption values are 86.0 and 88.1 kW hour, respectively, and their annual energy consumption is 31,384 and 32,173 kW hour, respectively. The annual cost saving for LS-PMSM is 2.45% higher than for IM. When the cable losses are taken into account the energy savings in relation to IM of both LS-SynRM and LS-PMSM raise to 3.6%.

The life cycle energy cost for 20 years of use is estimated in euros and is 108,651, 108,722 and 112,742 € including cable losses for LS-SynRM, LS-PMSM and IM, respectively. It can be seen that for LS-SynRM, the life cycle savings in energy costs against the IM excluding and including cable losses are relatively similar 3900 € and 4181 €, respectively. In case of using LS-PMSM, the savings increase almost double from 2626 € to 4020 € when the cable losses were taken into account. Thus, before making a choice of particular motor type and calculating the payback period, it is necessary to take into account not only the efficiency of the motors, but also the power factor and its effect on cable losses, as this can significantly affect the results of the feasibility study.

However, as we have already mentioned above, LS-PMSMs have a higher cost compared to IM and LS-SynRM due to the use of expensive rare-earth magnets in its design. In addition, the rare-earth elements processing from raw ore is associated with significant environmental damage [16]. Therefore, the LS-SynRM, which does not have permanent magnets and has a comparable cost to IM, is the most attractive pumping system in terms of energy saving even taking into account cable losses.

6. Conclusions

The present paper provides a comparative analysis of the energy consumption of electric motors of various types (LS-SynRM, LS-PMSM and IM) used as part of a 4 kW fixed-speed pump unit with throttle control. The comparison takes into account not only the efficiency of the motors at different pump loads, but also the effect of the motor power factor on cable losses in the power supply line.

It is shown that the life cycle energy cost savings for LS-SynRM compared to IM was 3900 € excluding cable losses. For LS-PMSM this number was 2626 €. However, when taking into account the cable losses, affected by the power factor and the total current of the motor, the savings for both LS-SynRM and LS-PMSM are almost the same: 4181 € and 4020 €, respectively.

Therefore, when selecting the motor type and calculating the payback period, it is necessary to take into account not only the efficiency of the motors, but also the motor power factor and its impact on the cable losses, as this can significantly affect the results of the feasibility study.

Thus, the use of LS-SynRM in the considered pump application provides lower cost of motor manufacturing and environmental friendliness, compared to LS-PMSM, and energy savings of more than 4000 € during the life cycle, compared to IM.

Based on this study, LS-SynRM can be suggested as the best alternative for the considered pump application since having the lowest energy consumption. In addition, LS-SynRM has the manufacturing cost comparable to IMs, while the cost of LS-PMSM is significantly higher, due to the presence of permanent magnets, the extraction of which is harmful to the environment.

The proposed method can be applied in the analysis of the energy consumption of other motor types used in pump stations, for example, line-start permanent magnet assisted synchronous reluctance motors (LS-PMaSynRM).

Author Contributions: Conceptual approach, V.K. and V.P.; data curation, D.G., S.O., and V.K.; software, S.O. and V.K.; calculations and modelling, S.O., V.D., V.K. and V.P.; writing of original draft, D.G., S.O., V.D., V.K. and V.P.; visualization, D.G., and V.K.; review and editing, D.G., S.O., V.D., V.K. and V.P. All authors have read and agreed to the published version of the manuscript.

Funding: This research received no external funding.

Acknowledgments: The authors thank the editors and reviewers for careful reading and constructive comments.

Conflicts of Interest: The authors declare no conflict of interest.

References

1. ABB Group. Process Performance IE3 Aluminum Motors, 50/60 Hz. The Product Range for Global Customers, Product Note. 2017. Available online: https://www.baldor.com/mvc/DownloadCenter/Files/9AKK107020 (accessed on 9 June 2020).
2. Almeida, A.; Ferreira, F.; Baoming, G. Beyond Induction Motors—Technology Trends to Move Up Efficiency. *IEEE Trans. Ind. Appl.* **2014**, *50*, 2103–2114. [CrossRef]
3. Commission Regulation (EU) 2019/1781 of 1 October 2019 Laying Down Ecodesign Requirements for Electric Motors and Variable Speed Drives Pursuant to Directive 2009/125/EC of the European Parliament and of the Council, Amending Regulation (EC) No 641/2009 with Regard to Ecodesign Requirements for Glandless Standalone Circulators and Glandless Circulators Integrated in Products and Repealing Commission Regulation (EC) No 640/2009, Document 32019R1781. Available online: https://eur-lex.europa.eu/legal-content/EN/TXT/PDF/?uri=CELEX:32019R1781&from=EN (accessed on 9 June 2020).
4. Doppelbauer, M. Update on IEC motor and converter standards. In Proceedings of the 6th International Motor Summit for Energy Efficiency powered by Impact Energy, Motor Summit 2016, Zurich, Switzerland, 11–12 October 2016.
5. Annex to the Communication from the Commission to the European Parliament, the European Council, the Council, the European Economic and Social Committee and the Committee of the Regions, The European Green Deal, Brussels, 11.12.2019, COM(2019) 640 Final. Available online: https://ec.europa.eu/info/sites/info/files/european-green-deal-communication-annex-roadmap_en.pdf (accessed on 9 June 2020).

6. WQuattro, Super Premium Efficiency Motor. WEG Group, Catalogue, Code 50025713, Revision 3 July 2017. Available online: https://static.weg.net/medias/downloadcenter/h01/hfc/WEG-w22-quattro-european-market-50025713-brochure-english-web.pdf (accessed on 9 June 2020).
7. Synchro Vert IE4 Super Premium Efficiency Motors, Bharat Bijlee, Catalogue. Available online: https://www.bharatbijlee.com/media/14228/synchrovert_catalogue.pdf (accessed on 5 May 2020).
8. Ismagilov, F.; Vavilov, V.; Gusakov, D. Line-Start Permanent Magnet Synchronous Motor for Aerospace Application. In Proceedings of the 2018 IEEE International Conference on Electrical Systems for Aircraft, Railway, Ship Propulsion and Road Vehicles and International Transportation Electrification Conference (ESARS-ITEC 2018), Nottingham, UK, 7–9 November 2018; pp. 1–5. [CrossRef]
9. Sorgdrager, J.; Wang, R.-J.; Grobler, A. Multi-Objective Design of a Line-Start PM Motor Using the Taguchi Method. *IEEE Trans. Ind. Appl.* **2018**, *54*, 4167–4176. [CrossRef]
10. Kurihara, K.; Rahman, M. High-Efficiency Line-Start Interior Permanent-Magnet Synchronous Motors. *IEEE Trans. Ind. Appl.* **2004**, *40*, 789–796. [CrossRef]
11. Azari, M.; Mirsalim, M. Line-start permanent-magnet motor synchronisation capability improvement using slotted solid rotor. *IET Electr. Power Appl.* **2013**, *7*, 462–469. [CrossRef]
12. Ghahfarokhi, M.; Aliabad, A.; Boroujeni, S.; Amiri, E.; Faradonbeh, V. Analytical modelling and optimization of line start LSPM synchronous motors. *IET Electr. Power Appl.* **2020**, *14*, 398–408. [CrossRef]
13. Sethupathi, P.; Senthilnathan, N. Comparative analysis of line-start permanent magnet synchronous motor and squirrel cage induction motor under customary power quality indices. *Electr. Eng.* **2020**, 1–11. [CrossRef]
14. Dent, P. Rare earth elements and permanent magnets. *J. Appl. Phys.* **2012**, *111*, 07A721. [CrossRef]
15. Goss, J.; Popescu, M.; Staton, D. A Comparison of an Interior Permanent Magnet and Copper Rotor Induction Motor in a Hybrid Electric Vehicle Application. In Proceedings of the IEEE International Electric Machines & Drives Conference, EMDC, Chicago, IL, USA, 12–15 May 2013; pp. 12–15. [CrossRef]
16. Lima, I.; Filho, W. *Rare Earth Industry*; Elsevier: Amsterdam, The Netherlands, 2015.
17. Kazakbaev, V.; Prakht, V.; Dmitrievskii, V.; Ibrahim, M.N.; Oshurbekov, S.; Sarapulov, S. Efficiency Analysis of Low Electric Power Drives Employing Induction and Synchronous Reluctance Motors in Pump Applications. *Energies* **2019**, *12*, 1144. [CrossRef]
18. Moghaddam, R.; Magnussen, F.; Sadarangani, C. Theoretical and Experimental Reevaluation of Synchronous Reluctance Machine. *IEEE Trans. Ind. Electron.* **2010**, *57*, 6–13. [CrossRef]
19. Kabir, M.; Husain, I. Application of a Multilayer AC Winding to Design Synchronous Reluctance Motors. *IEEE Trans. Ind. Appl.* **2018**, *54*, 5941–5953. [CrossRef]
20. Pellegrino, G.; Cupertino, F.; Gerada, C. Automatic Design of Synchronous Reluctance Motors Focusing on Barrier Shape Optimization. *IEEE Trans. Ind. Appl.* **2015**, *51*, 1465–1474. [CrossRef]
21. Kazakbaev, V.; Prakht, V.; Dmitrievskii, V.; Sarapulov, S.; Askerov, D. Comparison of power consumption of synchronous reluctance and induction motor drives in a 0.75 kW pump unit. In Proceedings of the 2017 International Siberian Conference on Control and Communications (SIBCON), Kazakhstan, Astana, 29–30 June 2017; pp. 1–6. [CrossRef]
22. ABB Group. Low Voltage IE4 Synchronous Reluctance Motor and Drive Package for Pump and Fan Applications, Catalogue. 2013. Available online: https://library.e.abb.com/public/21ee11b9fddfa677c1257bd500219300/Catalog_IE4_SynRM_EN_06-2013_9AKK105828_lowres.pdf (accessed on 9 June 2020).
23. Jia, S.; Zhang, P.; Liang, D.; Dai, M.; Liu, J. Design of IE4 Level Synchronous Reluctance Machines with Different Number of Poles. In Proceedings of the 2019 22nd International Conference on Electrical Machines and Systems (ICEMS), Harbin, China, 11–14 August 2019; pp. 1–5. [CrossRef]
24. Dmitriveskii, V.; Prakht, V.; Kazakbaev, V. IE5 Energy-Efficiency Class Synchronous Reluctance Motor with Fractional Slot Winding. *IEEE Trans. Ind. Appl.* **2019**, *55*, 4676–4684. [CrossRef]
25. Dmitrievskii, V.; Prakht, V.; Kazakbaev, V.; Oshurbekov, S.; Sokolov, I. Developing ultra premium efficiency (IE5 class) magnet-free synchronous reluctance motor. In Proceedings of the 2016 6th International Electric Drives Production Conference (EDPC), Nuremberg, Germany, 30 November–1 December 2016; pp. 2–7. [CrossRef]
26. REEL SuPremE®–The IE5* Magnet-Free Synchronous Reluctance Motor, KSB Company, Catalogue. 2019. Available online: https://www.ksb.com/blob/199782/10a90edea2eca4cb28afab3076d43df3/reel-supreme-motor-brochure-en-dow-data.pdf (accessed on 8 July 2020).

27. SIMOTICS Reluctance Motors with SINAMICS Converters, Siemens AG, Catalog. 2015. Available online: https://www.asberg.ru/upload/iblock/919/9195144e44c436fd94d723be0813ecfe.pdf (accessed on 9 June 2020).
28. Kersten, A.; Liu, Y.; Pehrman, D.; Thiringer, T. Rotor Design of Line-Start Synchronous Reluctance Machine with Round Bars. *IEEE Trans. Ind. Appl.* **2019**, *55*. [CrossRef]
29. Liu, H.; Lee, J. Optimum Design of an IE4 Line-Start Synchronous Reluctance Motor Considering Manufacturing Process Loss Effect. *IEEE Trans. Ind. Electron.* **2018**, *65*, 3104–3114. [CrossRef]
30. Kärkkäinen, H.; Aarniovuori, L.; Niemelä, M.; Pyrhönen, J.; Kolehmainen, J.; Känsäkangas, T.; Ikäheimo, J. Direct-On-Line Synchronous Reluctance Motor Efficiency Verification with Calorimetric Measurements. In Proceedings of the 2018 XIII International Conference on Electrical Machines (ICEM), Alexandroupoli, Greece, 3–6 September 2018; pp. 171–177. [CrossRef]
31. Hu, Y.; Chen, B.; Xiao, Y.; Shi, J.; Li, L. Study on the Influence of Design and Optimization of Rotor Bars on Parameters of a Line-Start Synchronous Reluctance Motor. *IEEE Trans. Ind. Appl.* **2020**, *56*, 1368–1376. [CrossRef]
32. Gamba, M.; Armando, E.; Pellegrino, G.; Vagati, A.; Janjic, B.; Schaab, J. Line-start synchronous reluctance motors: Design guidelines and testing via active inertia emulation. In Proceedings of the 2015 IEEE Energy Conversion Congress and Exposition (ECCE), Montreal, QC, Canada, 20–24 September 2015; pp. 4820–4827. [CrossRef]
33. Weatherization and Intergovernmental Programs Office, About the State Energy Program. Available online: https://www.energy.gov/eere/wipo/about-state-energy-program (accessed on 9 June 2020).
34. Federal Council Aims for a Climate-Neutral Switzerland by 2050. Available online: https://www.admin.ch/gov/en/start/documentation/media-releases.msg-id-76206.html (accessed on 9 June 2020).
35. Hopes of Limiting Global Warming? China and the Paris. Agreement on Climate Change. Available online: https://journals.openedition.org/chinaperspectives/6924?file=1 (accessed on 9 June 2020).
36. Japan Net Zero Energy Building Roadmap. Available online: http://www.nzeb.in/wp-content/uploads/2015/09/Japan-NZEB-Roadmap.pdf (accessed on 9 June 2020).
37. Ha, S.; Tae, S.; Kim, R. A Study on the Limitations of South Korea's National Roadmap for Greenhouse Gas Reduction by 2030 and Suggestions for Improvement. *Sustainability* **2019**, *11*, 3969. [CrossRef]
38. Shankar, V.; Umashankar, S.; Paramasivam, S.; Hanigovszki, N. A comprehensive review on energy efficiency enhancement initiatives in centrifugal pumping system. *Appl. Energy* **2016**, *181*, 495–513. [CrossRef]
39. Stoffel, B. Assessing the Energy Efficiency of Pumps and Pump Units. In *Background and Methodology*; Elsevier: Amsterdam, The Netherlands, 2015.
40. Goman, V.; Oshurbekov, S.; Kazakbaev, V.; Prakht, V.; Dmitrievskii, V. Energy Efficiency Analysis of Fixed-Speed Pump Drives with Various Types of Motors. *Appl. Sci.* **2019**, *9*, 5295. [CrossRef]
41. Ahonen, T.; Orozco, S.; Ahola, J.; Tolvanen, J. Effect of electric motor efficiency and sizing on the energy efficiency in pumping systems. In Proceedings of the 2016 18th European Conference on Power Electronics and Applications (EPE'16 ECCE Europe), Karlsruhe, Germany, 5–9 September 2016; pp. 1–9. [CrossRef]
42. Rhyn, P.; Pretorius, J. Utilising high and premium efficiency three phase motors with VFDs in a public water supply system. In Proceedings of the 2015 IEEE 5th International Conference on Power Engineering, Energy and Electrical Drives (POWERENG), Riga, Latvia, 11–13 May 2015; pp. 497–502. [CrossRef]
43. Brinner, T.; McCoy, R.; Kopecky, T. Induction versus Permanent-Magnet Motors for Electric Submersible Pump Field and Laboratory Comparisons. *IEEE Trans. Ind. Appl.* **2014**, *50*, 174–181. [CrossRef]
44. Ogunjuyigbe, A.; Jimoh, A.; Nicolae, D.; Obe, E. Analysis of synchronous reluctance machine with magnetically coupled three-phase windings and reactive power compensation. *IET Electr. Power Appl.* **2010**, *4*, 291–303. [CrossRef]
45. Obe, E. Steady-state performance of a line-start synchronous reluctance motor with capacitive assistance. *Electr. Power Syst. Res.* **2010**, *80*, 1240–1246. [CrossRef]
46. Gamba, M.; Pellegrino, G.; Vagati, A.; Villata, F. Design of a line-start synchronous reluctance motor. In Proceedings of the 2013 International Electric Machines & Drives Conference, Chicago, IL, USA, 12–15 May 2013; pp. 648–655. [CrossRef]
47. Wquattro 4 kW 4P 112M 3Ph 400/690 V 50 Hz IC411-TEFC-B3T. Product Details. Available online: https://www.weg.net/catalog/weg/RS/en/Electric-Motors/Special-Application-Motors/Permanent-Magnet-Motors/Line-Start-PM-Motors/Wquattro-4-kW-4P-112M-3Ph-400-690-V-50-Hz-IC411---TEFC---B3T/p/13009391 (accessed on 9 June 2020).

48. W22 IE3 4 kW 4P 112M 3Ph 220-240/380-415//460 V 50 Hz IC411-TEFC-B14T. Product Details. Available online: https://www.weg.net/catalog/weg/RS/en/Electric-Motors/Low-Voltage-IEC-Motors/General-Purpose-ODP-TEFC/Cast-Iron-TEFC-General-Purpose/W22---Cast-Iron-TEFC-General-Purpose/W22-IE3/W22-IE3-4-kW-4P-112M-3Ph-220-240-380-415-460-V-50-Hz-IC411---TEFC---B14T/p/14640586 (accessed on 9 June 2020).
49. Matine, A.; Bonnard, C.-H.; Blavette, A.; Bourguet, S.; Rongère, F.; Kovaltchouk, T.; Schaeffer, E. Optimal sizing of submarine cables from an electro-thermal perspective. In Proceedings of the European Wave and Tidal, Conference (EWTEC), Cork, Ireland, 27 August–1 September 2017; Available online: https://hal.archives-ouvertes.fr/hal-01754060 (accessed on 9 June 2020).
50. NM; NMS. Close Coupled Centrifugal Pumps with Flanged Connections; Catalogue; Calpeda. 2018. Available online: https://www.calpeda.com/system/pdf/catalogue_en_50hz.pdf (accessed on 5 May 2020).
51. IE4 DOLSynRM Concept Introduction, SynRM Benefits Available for Fixed Speed Applications. Tailored High Efficiency Motors for OEMs, Document Number: 9AKK106542 EN 04-2015. Available online: https://search.abb.com/library/Download.aspx?DocumentID=9AKK106542&LanguageCode=en&DocumentPartId=&Action=Launch (accessed on 9 June 2020).
52. Extended Product Approach for Pumps, Europump. 2014. Available online: http://europump.net/uploads/Extended%20Product%20Approach%20for%20Pumps%20-%20A%20Europump%20guide%20(27OCT2014).pdf (accessed on 9 June 2020).
53. Zhou, J.-H.; Chen, C.; Zhang, X.-W.; Chen, Y.-A. Reducing voltage energy-saving control method of induction motor. In Proceedings of the 2013 International Conference on Electrical Machines and Systems (ICEMS), Busan, Korea, 26–29 October 2013; pp. 2159–2162. [CrossRef]
54. Kirar, M.; Aginhotri, G. Cable sizing and effects of cable length on dynamic performance of induction motor. In Proceedings of the 2012 IEEE Fifth Power India Conference, Murthal, India, 19–22 December 2012; pp. 1–6. [CrossRef]
55. *Electrical Installations in Buildings—Part 5–52: Selection and Erection of Electrical Equipment—Wiring Systems, Is the IEC Standard Governing Cable Sizing*; IEC 60364-5-52; IEC: Geneva, Switzerland, 2009; Available online: https://webstore.iec.ch/publication/1878 (accessed on 8 July 2020).
56. Eurostat Data for the Industrial Consumers in Germany. Available online: http://ec.europa.eu/eurostat/statistics-explained/index.php/Electricity_price_statistics#Electricity_prices_for_industrial_consumers (accessed on 9 June 2020).
57. *Pump Life Cycle Costs: A Guide to LCC Analysis for Pumping Systems, Executive Summary*; Hydraulic Institute: Parsippany, NJ, USA; Europump: Brussels, Belgium; Office of Industrial Technologies Energy Efficiency and Renewable Energy, U.S. Department of Energy: Washington, DC, USA, 2001; pp. 1–19. Available online: https://www.energy.gov/sites/prod/files/2014/05/f16/pumplcc_1001.pdf (accessed on 8 July 2020).
58. Waghmode, L.; Sahasrabudhe, A. A comparative study of life cycle cost analysis of pumps. In Proceedings of the ASME 2010 International Design Engineering Technical Conferences and Computers and Information in Engineering Conference (ASME 2010), Montreal, QC, Canada, 15–18 August 2010; pp. 491–500. [CrossRef]

energies

Article

Simulation-Based Coyote Optimization Algorithm to Determine Gains of PI Controller for Enhancing the Performance of Solar PV Water-Pumping System

Jouda Arfaoui [1], Hegazy Rezk [2,3,*], Mujahed Al-Dhaifallah [4,*], Mohamed N. Ibrahim [5,6,7] and Mami Abdelkader [8]

1 National School of Engineering of Tunis, University of Tunis ELMANAR, BP 37, Tunis 1002, Tunisia; jouda.arfaoui@gmail.com
2 College of Engineering at Wadi Addawaser, Prince Sattam Bin Abdulaziz University, Al-Kharj 11991, Saudi Arabia
3 Electrical Engineering Deprtment, Faculty of Engineering, Minia University, Al Minya 61519, Egypt
4 Systems Engineering Department, King Fahd University of Petroleum & Minerals, Dhahran 31261, Saudi Arabia
5 Department of Electromechanical, Systems and Metal Engineering, Ghent University, 9000 Ghent, Belgium; m.nabil@ugent.be
6 FlandersMake@UGent—Corelab EEDT-MP, 3001 Leuven, Belgium
7 Electrical Engineering Department, Kafrelshiekh University, Kafr el-Sheikh 33511, Egypt
8 Department of Physics, Faculty of Sciences, University Tunis El Manar, BP 37, Tunis 1002, Tunisia; abdelkader.mami@gmail.com
* Correspondence: hr.hussien@psau.edu.sa (H.R.); mujahed@kfupm.edu.sa (M.A.-D.)

Received: 4 June 2020; Accepted: 27 August 2020; Published: 31 August 2020

Abstract: In this study, a simulation-based coyote optimization algorithm (COA) to identify the gains of PI to ameliorate the water-pumping system performance fed from the photovoltaic system is presented. The aim is to develop a stand-alone water-pumping system powered by solar energy, i.e., without the need of electric power from the utility grid. The voltage of the DC bus was adopted as a good candidate to guarantee the extraction of the maximum power under partial shading conditions. In such a system, two proportional-integral (PI) controllers, at least, are necessary. The adjustment of (Proportional-Integral) controllers are always carried out by classical and tiresome trials and errors techniques which becomes a hard task and time-consuming. In order to overcome this problem, an optimization problem was reformulated and modeled under functional time-domain constraints, aiming at tuning these decision variables. For achieving the desired operational characteristics of the PV water-pumping system for both rotor speed and DC-link voltage, simultaneously, the proposed COA algorithm is adopted. It is carried out through resolving a multiobjective optimization problem employing the weighted-sum technique. Inspired on the *Canis latrans* species, the COA algorithm is successfully investigated to resolve such a problem by taking into account some constraints in terms of time-domain performance as well as producing the maximum power from the photovoltaic generation system. To assess the efficiency of the suggested COA method, the classical Ziegler–Nichols and trial–error tuning methods for the DC-link voltage and rotor speed dynamics, were compared. The main outcomes ensured the effectiveness and superiority of the COA algorithm. Compared to the other reported techniques, it is superior in terms of convergence rapidity and solution qualities.

Keywords: simulation-based optimization; coyote optimization algorithm; water pumping; energy efficiency

1. Introduction

Thanks to the huge availability of the sun supply in comparison with other energy sources, the solar PV systems seem one of the most encouraging applications of renewable energy today and in the forthcoming years. Moreover, the cost of the PV modules is reducing increasing a result of the progress in the manufacturing technology of the solar modules [1]. Standalone and grid-connected systems are the two main types of solar systems. Indeed, Sun availability is taking as one of the most important motivation for depending on the solar system as a promising solution. As being a type of renewable energy resources, the solar PV has gained a growing importance for uses in the electric power domain which has several advantages like (1) being harmless energy, (2) being suitable for isolated sites and (3) being cheap and easy maintenance.

Indeed, the efficiency and cost of the PV pumping system are not attractive yet to spread widely, but, many literature works are devoted to improving the total effectiveness and reducing the total cost of the solar pumping generation system [1,2].

A major drawback of using this sort of vector control lies in the fact that the effectiveness of the water-pumping system is in the heavy dependence on the appropriate adjustment of the PI parameters. Commonly, the set of these parameters is based on the classical and tiresome trials and error methods [3–6].

Adjustment method turned out to be complex and time-consuming. Moreover, the conventional PI tuning approach such as the symmetrical optimum [7], Ziegler–Nichols [8], Tyreus–Luyben [9] and Cohen–Coon [10] techniques requires that the designer has a good insight about the characteristics and dynamics of the controlled system. Hence, for the sake of overcoming these drawbacks, a systematic process to identify best gains of PI regulators has been provided in [3] and the modern optimization is considered as an auspicious solution.

In [11] the authors considered two optimization metaheuristics: Ant colony optimization (ACO) and differential evolution (DE). These strategies are reported to provide an optimized adjustment (PI) regulators in the direct torque control–space vector modulation (DTC-SVM) control loops, like rotor speed, electromagnetic torque, stator flux linkage and estimation of the linkage stator flux. Compared to other metaheuristics methods, those considered approaches require few tuning parameters to provide a fast convergence rate. Simulation and practical experimental on three phase induction motor (TIM) controlled by DTC-SVM are carried out. On the other hand, classical procedures such as frequency and root locus have been applied.

In [12], the study reported a novel optimization algorithm that combines all these three techniques: genetic algorithm (GA), artificial neural networks (ANN) and fuzzy logic, which was named as GNFPID. This technique aims to outperform the performance of the used PID controller. In [13], a genetic algorithm is applied as the adequate adjustment method of PI controllers applied to DTC drive for doubly fed asynchronous machine. Following in the same strategy, particle-swarm optimization (PSO), bat algorithm (BA) and genetic algorithm (GA) are considered for the adequate tuning of the gains of PI for brushless DC motor in [14].

Metaheuristics are an iterative processes for solving different optimization problems by taking into account simple trial operations [15,16]. Particle-swarm optimization [17], genetic algorithm [18], Imperialist competitive algorithm [19], ant colony optimization [20,21], differential evolution [22,23] are commonly treated. In [24], water cycle algorithm was adopted to adjust the parameters of PI controller for speed control and reactive power control loops in RSC (rotor side converter) and DC-link regulation voltage loop in GSC (grid side converter).

In the same way, [25] has referred to a thermal exchange optimization (TEO), this metaheuristic method was applied for the tuning of PI controller gains for the external loops in the conventional vector control strategy of a doubly fed induction generator based on wind turbine system. For assessing the efficiency and superiority of the proposed control strategy, other well-known metaheuristic procedures such as grasshopper optimization algorithm (GOA), particle-swarm optimization (PSO), water cycle algorithm (WCA) and harmony search algorithm (HSA), are taken into account for comparison purposes.

To best of our knowledge, for first time, the coyote optimization (COA) algorithm is used to identify the parameters of PI controller to improve the performance of solar PV water-pumping system under partial shading condition. Therefore, in this study, COA algorithm is investigated and further, to prove the superiority of the suggested COA algorithm, the obtained results are compared with the classical Ziegler–Nichols method, trial–error tuning method and PSO algorithm.

The remainder of this study is arranged as follows: the mathematical model of the solar photovoltaic water-pumping system (SPWPS) is described in Section 2. Section 3 is devoted to the formulation of the outer-loops PI controllers' design problem given as a constrained optimization problem. The proposed COA method is given in detail, aimed to resolve such a problem. In Section 4, the demonstrative results of the COA-tuned PI controllers tuning are carried out. Concluding remarks are summarized in Section 5.

2. Modeling of the Solar Photovoltaic Water-Pumping System

Referring to the solar pumping system, several control methods were developed to provide an effective functioning. These include speed control, distinct maximum power point tracking (MPPT) techniques, voltage control and frequency regulation under load variation [26].

This part studies and simulates a new structure of a solar pumping system. The overall structure of the proposed PV pumping system is described, as well as its various subsystems.

Vector control also called field-oriented control, is a method that permits independent control of the flux and torque components. It is equipped with high-performance drives. In this section, the interest is focused on the type of control.

The proposed system includes the following parts: PV panel, three-phase voltage inverter, DC–DC boost converter and induction engine-driven pump unit.

A schematic diagram of the new structure including indirect field oriented control (IFOC) and MPPT strategies, is illustrated in Figure 1. Concerning the control of this sort of system, different techniques were considered for effective functioning in terms of speed control, voltage control, distinct MPPT techniques and frequency regulation under load variations. The most important parts of this system are indicated in Figure 1 and are recapitulated as follows:

(1) Shaded PV modules represent the power supply source of the induction engine via a three-phase voltage source inverter (VSI);
(2) A hysteresis pulse-width modulation (PWM) based current control;
(3) A boost converter for forcing the PV panel to operate at the MPP under different partial shading conditions [27];
(4) Flux weakening component is required to produce the reference current (I^*_{ds}) and the speed controller output gives (I^*_{qs});
(5) Vector control of asynchronous machine for driving centrifugal pump [28].

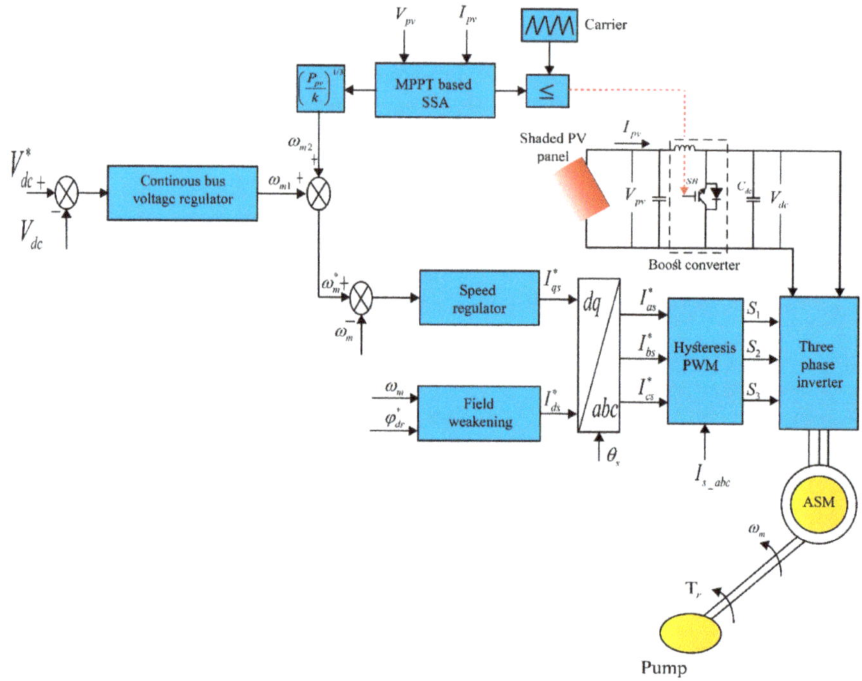

Figure 1. Scheme diagram of the proposed system.

The reference speed is a measure of the generated PV power. The PI controllers' design and the suitable values of the gains regulators (K_{ps}, K_{is}, K_{pdc} and K_{idc}) are typically tuned employing trials and errors technique, as mentioned before. Regarding the control design stage of VSI, this hard and challenging task turns into more difficult and time-consuming. A recent global meta-heuristic algorithm is an effective solution to tackle this matter.

2.1. Model of the Induction Machine

In this context, the Park transformation is adopted when modeling the asynchronous machine to omit the inductance variation with time [29].

$$V_{ds} = R_s I_{ds} + \sigma L_s \frac{dI_{ds}}{dt} + \frac{L_m}{L_r}\frac{d\varphi_{dr}}{dt} - \omega_s \sigma L_s I_{qs} \quad (1)$$

$$V_{qs} = R_s I_{qs} + \sigma L_s \frac{dI_{qs}}{dt} + \omega_s \frac{L_m}{L_r}\varphi_{dr} + \omega_s \sigma L_s I_{ds} \quad (2)$$

$$V_{dr} = 0 = \frac{R_r}{L_r}\varphi_{dr} + \frac{d\varphi_{dr}}{dt} - \frac{L_m}{L_r} R_r I_{ds} \quad (3)$$

$$V_{qr} = 0 = \omega_{sl}\varphi_{dr} - \frac{L_m}{L_r} R_r I_{qs} \quad (4)$$

where

$$\sigma = 1 - \frac{{L_m}^2}{L_s L_r}$$

V_{ds}, V_{qs}, V_{dr} and V_{qr} indicate the stator and the rotor voltages, respectively.

I_{ds} and I_{qs} represent the stator currents.

$$\varphi_{dr} = L_m I_{mdr} \tag{5}$$

$$I_{ds} = I_{mdr} + \frac{L_r}{R_r}\frac{dI_{mdr}}{dt} \tag{6}$$

$$I_{ds} = (1 + p{\cdot}\tau_r) I_{mdr} \tag{7}$$

where

$$\tau_r = \frac{L_r}{R_r}$$

In steady states conditions

$$I_{ds} = I_{mdr} \tag{8}$$

The instantaneous value of the angle θ_s is determined by this equation:

$$\theta_s = \int \omega_s dt \tag{9}$$

$$\theta_s = \int \omega_s dt = \int (\omega_{sl} + \omega_r) dt \tag{10}$$

The sliding speed ω_{sl} can be calculated using this equation [29]:

$$\omega_{sl} = \frac{L_m}{L_r}\frac{R_r}{\varphi_{dr}} I_{qs} = \frac{L_m I_{qs}}{\tau_r \varphi_{dr}} = \frac{I_{qs}}{\tau_r I_{mdr}} \tag{11}$$

In steady states conditions ($I_{mdr} = I_{ds}$) and thus

$$\omega_{sl} = \frac{I_{qs}}{\tau_r I_{ds}} \tag{12}$$

Equations (13) and (14) are needed to determine the reference flux and reference stator current, respectively [29]:

$$\varphi^*_{dr} = \begin{cases} \varphi_{rn} & si\ |\omega_m| \le \omega_{mn} \\ \frac{\varphi_{rn}\omega_{mn}}{|\omega_m|} & si\ |\omega_m| > \omega_{mn} \end{cases} \tag{13}$$

where φ_{rn} denotes nominal value of flux.

ω_{mn} denotes nominal value of mechanical speed

$$I^*_{ds} = \frac{\varphi^*_{dr}}{L_m} \tag{14}$$

L_m: is the mutual inductance

$$I^*_{qs} = T^*_e / \frac{3}{2}\frac{PL_m}{2L_r}\varphi^*_{dr} \tag{15}$$

The correlation process between the torque and the speed of the pump model is described as follows [30]:

$$T_r = a_1 + a_2{\cdot}{\omega_m}^2 \tag{16}$$

where

a_1: constant;

a_2: constant;

T_r: Torque of pump;

w_m: motor speed.

The power of the pump has a cubic relationship with the motor speed w_m, thus driving the centrifugal pumps [31,32].

To evaluate one of the reference speed components, an affinity law shall be applied for this purpose. Under different solar irradiation, the induction machine operates and seeks to determine the centrifugal pump flow rates. Moreover, to optimize the produced PV power, a MPPT control was investigated. It enables the extraction of the global MPP. The generated power serves to determine the first component of the reference speed as follows [33]:

$$\omega_{m1} = \left(\frac{P_{pv}}{k}\right)^{1/3} \tag{17}$$

The PV array power is converted in terms of speed via the constant k.

The second component of the reference speed is evaluated by means of the DC-link voltage controller. In this system, a comparison between the measured bus voltage (V_{dc}) and the desired bus voltage (V^*_{dc}), is carried out, which conducts to a voltage error (ΔV_{dc}), described as [34]:

$$\Delta V_{dc}(n) = V^*_{dc}(n) - V_{dc}(n) \tag{18}$$

The output speed can be obtained by the following equation [34]:

$$\omega_{m2}(n) = \omega_{m2}(n-1) + K_{pdc}\{\Delta V_{dc}(n) - \Delta V_{dc}(n-1)\} + K_{idc} \cdot \Delta V_{dc}(n)\} \tag{19}$$

where K_{pdc} and K_{idc} are, respectively the proportional and integral parameters of the DC-link-voltage controller.

The reference speed of the induction motor is obtained according to the following equation [33]:

$$\omega^*_m = \omega_{m1} + \omega_{m2} \tag{20}$$

where ω^*_m is the reference speed and w_{m1} is the first part of reference speed, determined by means of Equation (17).

w_{m2} is the second part of reference speed, determined by means of Equation (19).

2.2. Pump Model

The mechanical model of the induction motor actuating the centrifugal pump is provided by the following formula [35]:

$$T_e - T_r - f\omega_m = J\frac{d\omega_m}{dt} \tag{21}$$

T_r and T_e represent the load torque and the electromagnetic torque, respectively, J is the moment of inertia and f is the damping coefficient which links the friction torque C_f to the motor speed by relation ($C_f = -f\omega_m$).

The following equation used to represent the pump torque [33]:

$$T_r = A_p\omega^2_m \tag{22}$$

where A_p denotes torque constant and estimates as following [33]:

$$A_p = \frac{P_n}{\omega^3_{mn}} \tag{23}$$

where P_n and w_{mn} indicate rated power and nominal value of mechanical speed, respectively.

2.3. Three-Phase Voltage Inverter

To achieve an optimum control of the power supply between the PV source and the asynchronous engine, a voltage source inverter is required as a crucial part. The employed inverter consists of three independent arms, as illustrated in Figure 2. DC–AC converters are dealing with in this section, which already applied in the solar energy sector as well as in the three-phase voltage inverters. The state of the switches (k_n) has a considerable impact on the inverter modeling. The produced voltages are provided by the following matrix form [36]:

$$\begin{bmatrix} V_1 \\ V_2 \\ V_3 \end{bmatrix} = \frac{V_{pv}}{3}\begin{bmatrix} 2 & -1 & -1 \\ -1 & 2 & -1 \\ -1 & -1 & 2 \end{bmatrix}\begin{bmatrix} S_1 \\ S_2 \\ S_3 \end{bmatrix} \tag{24}$$

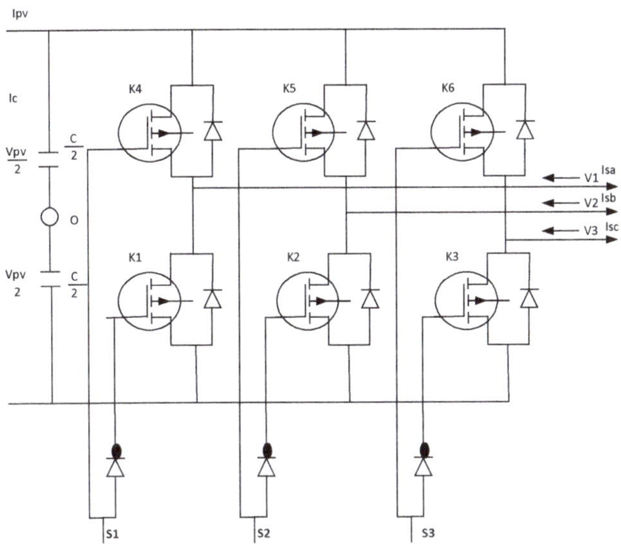

Figure 2. Three-phase voltage inverter.

The current provided by the DC source obeys to;

$$I_{pv} = S_1 I_{sa} + S_2 I_{sb} + S_3 I_{sc} + I_c \tag{25}$$

A PWM hysteresis strategy is employed to control these power converters.

2.4. DC–DC Boost Converter

An ideal scheme of the DC–DC boost converter is illustrated in Figure 3. The relation of proportionality between the input voltage (E) and the output voltage, is described as follows:

$$V_s = \frac{E}{1-u} \tag{26}$$

where u denotes the duty cycle.

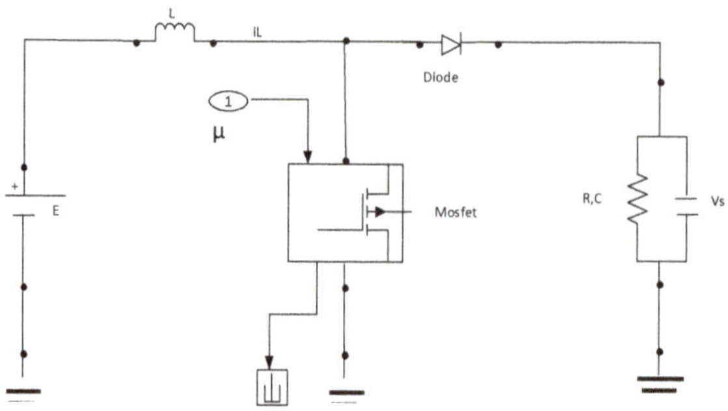

Figure 3. Scheme of DC converter.

The DC converter can be integrated with MPPT strategy to extract the maximum power from the PV array.

2.5. Hysteresis Based PWM Current Controller

In most studies, the PWM technique is the most common strategy since it perfectly adapts with the use of the voltage inverter. It is recommended also in case the vector control is adopted.

The principle of this command as presented in Figure 4 is to generate a PWM signal based directly on the variable to be controlled, through decisions based on the on–off control. In addition, it consists of maintaining the variation of the measuring currents within a band at a self-adjusting width, centered on the reference currents. The hysteresis comparator of any PWM structure receives as an input signal: the difference between the reference and measurement currents in order to produce at the output the control signal for the power switches [36]. In fact, this control is simple to operate as it does not require precise knowledge of the machine but suffers from the lack of control of the switching frequency.

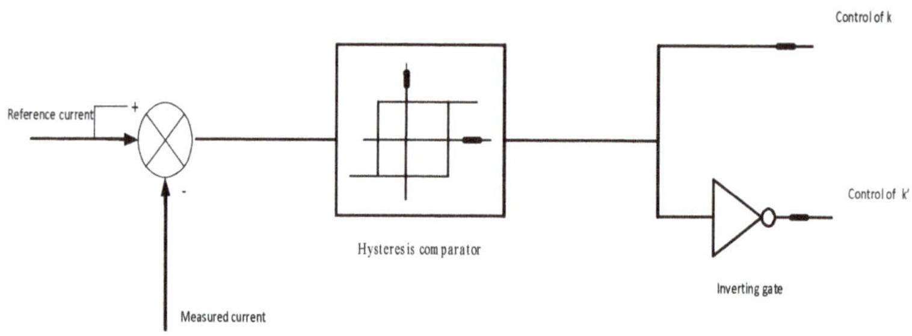

Figure 4. Schematic diagram of the hysteresis pulse-width modulation (PWM) control.

3. COA Design for Optimal PI Parameters Tuning

The coyote optimization algorithm (COA) was recently proposed by Pierezan and Coelho [37]. The core idea of COA optimizer is based on *Canis latrans* species that reside mainly North America [37]. The algorithm is adapted to consider the social organization of the agents called coyotes and it has been serving with a different algorithmic construction. The greatest advantage of this method is to preserve a balance between the exploitation and exploration mechanisms during that optimizing process. By contrast with the gray wolf optimization (GWO), the COA does not center on how these dominant norms have been followed by these animals and the social hierarchy. Additionally, the COA depends not only on the hunting preys which take place in the GWO, but also on the social structure and regular experience interchange which are carried out by the coyotes. They are distinguished by cooperative functionalities as they head toward the prey in the close chain while they have a strong sense of smell which makes it possible to identify the location of the prey. Regarding the hunting process, coyotes attack in groups, this action forces the agents to update their positions to improve them. When coyotes' striking their opponents, they are well-prepared with a probability of threat and move away from their current position by an excessive random distance. Further, the COA method is initialized by an initializing population of coyotes whose size can be defined as the multiplication of packs and coyotes. The number of agents called coyotes in each pack is supposed to be equal and constant. In that regard, to facilitate the users' understanding, each coyote can be considered as a candidate solution for the optimization problem and its social condition presents the fitness function. The COA optimization algorithm started by a population of coyotes that are created randomly within the search space. More details about physical meaning and mathematical representation can be found in Ref [37,38]. Figure 5 presents the optimization process of COA.

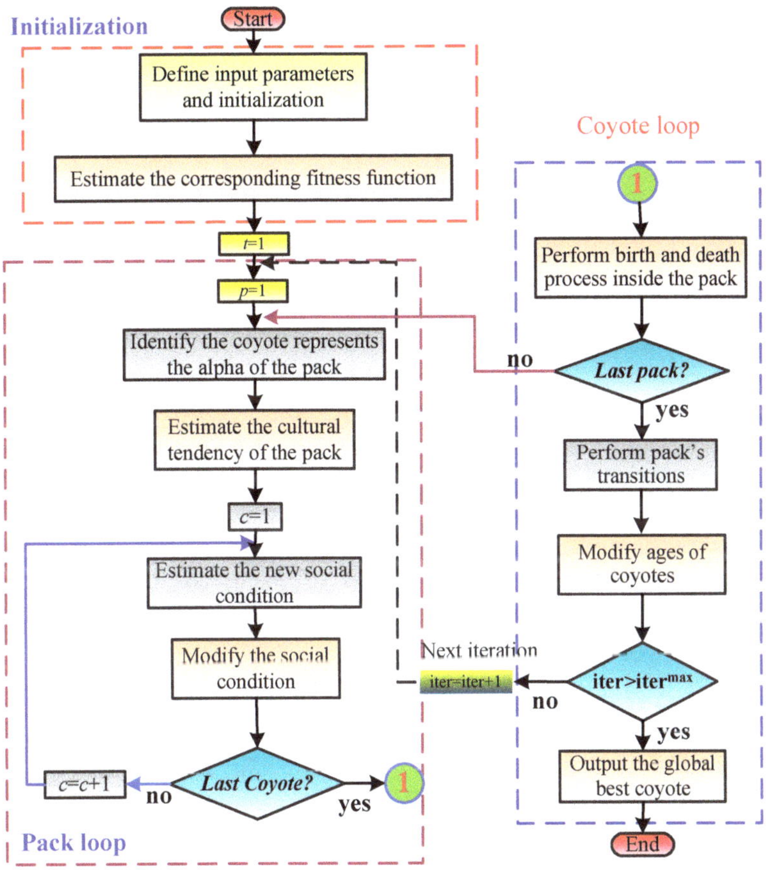

Figure 5. Optimization process of coyote optimization algorithm (COA).

Regarding the PI design control, the suitable parameters of this type of controllers are usually tuned by trials and error techniques [3–6]. This non-systematic process turns into more complicated and into more time-consuming, particularly for the complex systems' design. Accordingly, the idea of transforming the identification process of these parameters (K_{ps}, K_{is}, K_{pdc} and K_{idc}) to an optimization problem, is a promising alternative. By means of some advanced meta-heuristic algorithms, such optimization problems can be effectively managed [3]. Within the framework provided, two interactive PI regulators for speed and DC-link voltage loops are being introduced. These PI regulators are demanded to achieve an optimized value of this control using such COA algorithm. Figure 6 provides the suggested tuning process of PI controllers for the solar photovoltaic water-pumping system (SPVWPS) based on optimization approach.

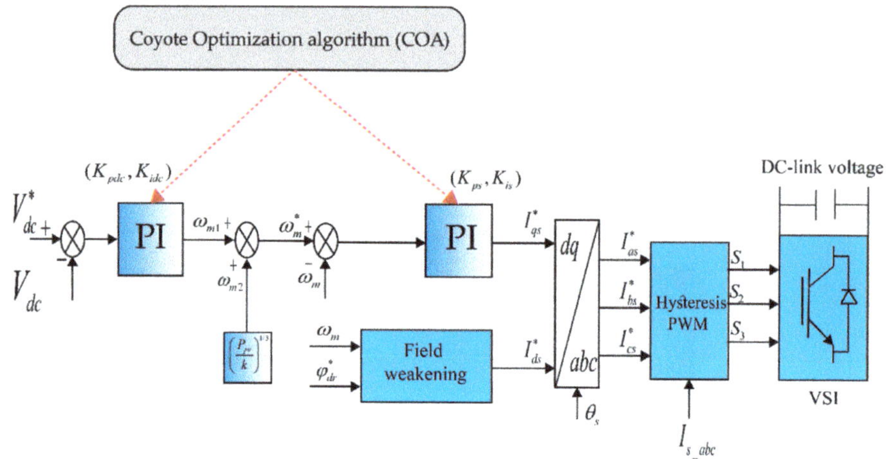

Figure 6. Proposed metaheuristics-tuned proportional-integral (PI) controllers for the solar photovoltaic water-pumping system (SPVWPS).

During the optimization process, the gains of PI controllers for the DC-link voltage and speed loops are considered as the decision variables of the problem, these parameters are denoted as follows:

$$x = \left[K_{pdc}, K_{idc}, K_{ps}, K_{is}\right]^T \in \mathbb{S} \subseteq \mathbb{R}_+^4 \tag{27}$$

where K_{pdc} and K_{idc} indicate the proportional and integral gains of DC-link regulator and K_{ps} and K_{is} denote the proportional and integral gains of speed regulator.

Under time-domain, such constraints like rise time, maximum overshoot, settling time and steady-state error of the system step-response, the performance criterion is minimized according to the suitable parameters of the problem [3].

The tuning process based on optimization is defined as follows:

$$\begin{cases} \text{minimize } g_i(\boldsymbol{x}) \\ \boldsymbol{x} = \left[K_{pdc}, K_{idc}, K_{ps}, K_{is}\right]^T \in \mathbb{S} \subseteq \mathbb{R}_+^4 \\ \text{subject to :} \\ h_1(\boldsymbol{x}) = \delta_{Vdc} - \delta_{Vdc}^{\max} \leq 0 \\ h_2(\boldsymbol{x}) = \delta_{speed} - \delta_{speed}^{\max} \leq 0 \end{cases} \tag{28}$$

In formulating the optimization problem, cost function are outlined as follows:

$$g_{IAE,i}(x) = \int_0^\infty \left|e_i(x,t)\right| dt \tag{29}$$

Which are adopted the Integral of Absolute error (IAE) and maximum overshoot (MO) criteria, $\mathbb{S} = \left\{x \in \mathbb{R}_+^4, x_{low} \leq x \leq x_{up}\right\}$ is the initial search space.

Under such circumstances and to show superior performance, some constraints had to be considered in the optimization framework, $h_j : \mathbb{R}^4_+ \to \mathbb{R}, j \in \{1, 2\}$ presented the inequality constraints which are considered for this optimization process.

The terms δ_{dc} and δ_{speed} are the overshoot of the DC-link voltage and rotor speed, respectively. The optimization process described in Equation (28) is considered as a multi-objective optimization type; an aggregate of all costs functions to form one single objective function is determined by means of an aggregation function [3].

$$g(x) = \sum_{i=1}^{2} \omega_i g_i(x) \tag{30}$$

where $w_i > 0$, it denotes the weighting coefficients of the aggregation cost function. The objective function befitting for COA-tuned PI regulators for SPVWPS under shading condition, reformulated as follows:

$$g(x) = \int_0^{+\infty} \left(\omega_1 \left|w_r^* - w_r\right| + \omega_2 \left|v_{dc}^* - v_{dc}\right|\right) dt \tag{31}$$

4. Results and Discussion

The considered system in this research work contains a three-phase induction motor drive of 2200 W, 230 V, used to feed the water-pumping system. The main source is PV array with rating of 2400 W. The PV array composed of 234 PV module. The number of series PV modules interconnected is series is 18 modules whereas number of parallel string is 13 strings. The open circuit voltage and short circuit current of PV modules are 21.6 V and 0.64 A, respectively whereas the voltage and current at maximum power point are 17.6 V and 0.58 A. The case study considered the partial shading condition, three different level of the solar radiation; 1.0 kW/m^2, 0.8 kW/m^2 and 0.5 kW/m^2 are subjected to the PV array.

In order to tackle the problem of time-consuming, the PI control tune by COA is performed, which further highlights the contribution of the proposed adjustment approach based on metaheuristic versus the given classical Ziegler–Nichols and Trial and Error method.

Table 1 provides a summary of the obtained gains of the PI controllers for the classical based methods and the proposed optimization meta-heuristic called the coyote optimization algorithm (COA). Furthermore, a comparison is carried out between the obtained results for the proposed COA-tuned PI controllers and those obtained by the well-known Ziegler–Nichols and Trial and Error method for the PI controllers' design of DC-link voltage (Table 2 and Figure 7).

Table 1. Optimized PI controller gains.

	Algorithms		
Parameter	**Trial and Error**	**Ziegler–Nichols**	**COA**
K_{ps}	0.2499	4.78577	1.8706
K_{is}	16.9730	229.6941	3.8149
K_{pdc}	1	0.3052	2.7748
K_{idc}	500	62.2047	1.3374

Table 2. Output performance under different methods.

Methods	δ (%)	*tr* (sec)	*ts* (sec)	*Ess*
Trial–Error	20	0.1238	0.26	0.006
Ziegler–Nichols	5.1	0.0279	0.2	0.004
PSO-Method	0	0.0332	0.987	0.03
COA-Method	0	0.015	0.151	0.001

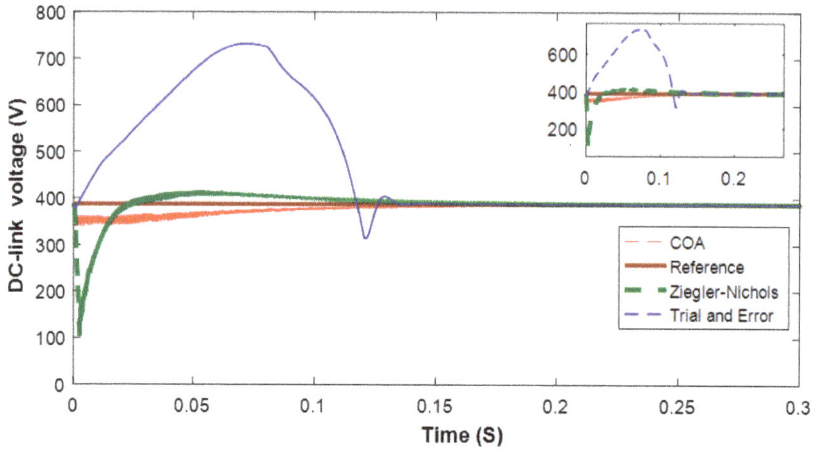

Figure 7. DC-link voltage under different methods.

According to these results, COA-tuned PI controllers with the IAE criteria, outperforms these considered classical based methods in terms of settling time, overshoot and steady-state error indices. Indeed, the DC-link voltage performance using COA for designing the PI controller has a negligible steady-state error and also a negligible overshoot and it provides a shorter rise time than that in the case of a conventional adjustment method which presents a higher overshoot at the transient operation. For such reformulated PI design issues and in terms of offering the quality of solutions and non-premature convergence as well as fastness, the COA algorithm outperforms all other methods. To reinforce the effectiveness of the proposed COA method, the homologous PSO algorithm is considered for comparison purposes.

Figure 8 represents the dynamic response of the DC-link voltage based on the PI parameters design by COA and PSO. It can be observed from this figure, that COA can maintain the DC-link voltage to constant value faster than that of the other meta-heuristic (PSO). This confirms the higher global convergence of COA due to the impact of exploration and exploitation mechanisms that are developed by that algorithm. From Figure 8, the tracking dynamic indicates a considerable oscillating using the PSO algorithm for adjustment PI controller design and the COA tuned PI controller enables us to reach rapidly the steady-state. To confirm the reliability of COA compared to PSO, both algorithms executed 30 times. The population size and maximum of iterations are 20 and 50, respectively. The results of statistical comparison are presented in Table 3.

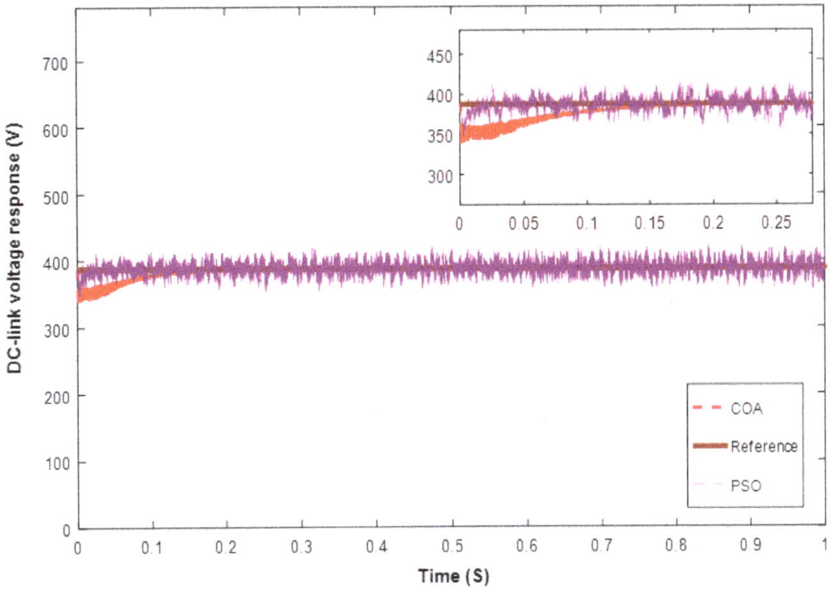

Figure 8. DC-link voltage responses using particle-swarm optimization (PSO) and COA.

Table 3. Statistical comparison between PSO and COA.

Algorithms	Best	Mean	Worst	Median	STD
PSO	3.4528	6.5660	9.8254	6.2777	2.23
COA	2.1696	5.3835	8.5281	5.8741	2.14

The detailed performance (power, speed, DC-link voltage, load torque, electromagnetic torque, etc. . . . .) of the PV generation system, obtained with COA algorithm used for tuning PI controllers design of the DC-link voltage and rotor speed, are illustrated in Figure 9.

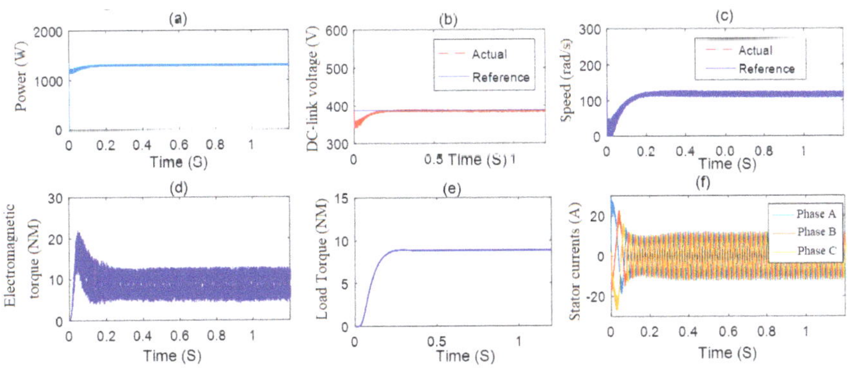

Figure 9. Performance of SPVWPS under partial shading condition using COA algorithm tuned PI controller of the DC link voltage and rotor speed. (**a**) power; (**b**) DC voltage; (**c**) speed; (**d**) electromagnetic torque; (**e**) load torque and (**f**) stator currents.

Considering Figure 9, it can be observed that the desired and the detected values of the rotor speed converge simultaneously to the nominal value that demonstrate the ameliorated efficiency of the speed control. Considering the heavy dependence between the rotor speed components and the load torque, the water spilling out of the pump was reduced due to sudden torque variation in the suggested regulation method. Figure 9c illustrates the rotor speed behavior of the PV system due to the increased proportionally of the load torque, however, it achieves steady-state shortly at 0.2 s.

However, the proposed system uses an MPPT algorithm that provides the reference speed as an output signal, to be fed into indirect-field oriented control (IFOC) strategy. It is worth noting that the reference and measured values of the DC bus voltage converge towards the desired value, which has demonstrated an enhanced efficiency of the voltage regulation. The reference value (set-point) of the DC bus voltage must be constant and its value was selected to 388 V. The outer PI controller loop is adopted to keep the DC bus voltage constant. The actual value of the DC bus voltage is the value that is measured by the voltage sensor.

In other words, this Set point (reference speed), which must be applied to the machine, comes from a new maximum power tracking algorithm based on the PSO technique (PSO-MPPT) and a pump affinity law (Equation (16)). This process results in an optimization of photovoltaic power under partial shading conditions. Further, it can be seen that the load torque and the rotor speed follow the PV power. This is found in Figure 9. It was found that the inaccuracy of the speed has an important effect on the efficiency and reliability of the pump drive (The relationship of proportionality was indicated in Equation (16)).

In sum, the COA algorithm can identify the gains of PI controllers of DC link voltage and rotor speed in a reduced computational time and under functional constraints compared to the reported classical techniques which usually provide local solutions for the control problem with constraints.

5. Conclusions

The optimal gains of PI controllers were determined using coyote optimization algorithm (COA) in order to enhance the effectiveness of the solar photovoltaic water-pumping system. In fact, the controllers' adjustment is reformulated as a constraint optimization problem under the functional time domain. A comparative study of the dynamic response of the DC-link voltage and rotor speed for the COA-tuning PI controllers is carried out with the well-known classical Ziegler–Nichols and Trial and Error tuning procedures. Similarly, the proposed meta-heuristic COA succeeds in the tune of PI controllers for DC-link voltage and rotor speed more effectively than the considered meta-heuristic PSO algorithm. The COA provides good responses of the DC-link voltage and rotor speed, during the transient and steady state cycle. The main outcomes confirmed the performance superiority of the developed COA-tuned PI controllers, in terms of the rapidity of convergence and solutions quality. To reduce the cost of the controller, a senseless-MPPT will be considered in the future work.

Author Contributions: Conceptualization, J.A., M.N.I. and H.R.; methodology, J.A., M.N.I and H.R.; software, J.A., H.R. and M.A.-D.; formal analysis, J.A., M.N.I., M.A. and H.R.; investigation, J.A., M.A.-D., M.A.; resources, J.A, M.A.-D. and H.R.; writing—original draft preparation, J.A., M.N.I., M.A.-D., M.A. and H.R.; writing—review and editing, J.A., M.N.I., M.A.-D., M.A. and H.R. All authors contributed equally to this work. All authors have read and agreed to the published version of the manuscript.

Funding: This research received no external funding.

Conflicts of Interest: The authors declare no conflict of interest.Abbreviations

Abbreviations

ACO	Ant colony optimization
ANN	Artificial neural network
BA	Bat algorithm
COA	Coyote optimization algorithm
DC	Direct current

AC	Alternating current
DE	Differential evolution
DTC-SVM	Direct torque control–space vector modulation
GA	Genetic algorithm
GNFPID	Genetic neural fuzzy proportional integral derivative
GOA	Grasshopper optimization algorithm
GSC	Grid side converter
GWO	Gray wolf optimization
HSA	Harmony search algorithm
IAE	Integral absolute error
IFOC	Indirect field oriented control
MO	Maximum overshoot
MPP	Maximum power point
MPPT	Maximum power point tracking
PI	Proportional integral
PID	Proportional integral derivative
PSO	Particle-swarm optimization
PV	Photovoltaic
PWM	Pulse-width modulation
TIM	Three phase induction motor
RSC	Rotor side converter
SPVWPS	Solar photovoltaic water-pumping system
TEO	Thermal exchange optimization
VSI	Voltage source inverter
WCA	Water cycle algorithm
Symbols	
f	Damping coefficient
I_{ds},I_{qs}	Stator currents
L_m	magnetizing inductance
L_r	rotor inductance
P_n	Rated power
P_{pv}	PV power
T_e	Electromagnetic torque
T_r	Load torque
V_{dc}	DC-link voltage
V_{dr}, V_{qr}	Rotor voltages
V_{ds}, V_{qs}	Stator voltages
ω_m	Motor speed
ω_{mn}	Nominal value of mechanical speed
ω_{sl}	Sliding speed
φ_{dr}	Rotor flux
φ_{rn}	Nominal value of flux

References

1. Ibrahim, M.N.; Rezk, H.; Al-Dhaifallah, M.; Sergeant, P. Solar Array Fed Synchronous Reluctance Motor Driven Water Pump: An Improved Performance under Partial Shading Conditions. *IEEE Access* **2019**, *7*, 77100–77115. [CrossRef]
2. Ibrahim, M.N.; Rezk, H.; Al-Dahifallah, M.; Sergeant, P. Hybrid Photovoltaic-Thermoelectric Generator Powered Synchronous Reluctance Motor for Pumping Applications. *IEEE Access* **2019**, *7*, 146979–146988. [CrossRef]
3. Bouallègue, S.; Haggège, J.; Ayadi, M.; Benrejeb, M. PID-type fuzzy logic controller tuning based on particle swarm optimization. *Eng. Appl. Artif. Intell.* **2012**, *25*, 484–493. [CrossRef]

4. Hu, H.; Hu, Q.; Lu, Z.; Xu, D. Optimal PID controller design in PMSM servo system via particle swarm optimization. In Proceedings of the 31st Annual Conference of IEEE Industrial Electronics Society, IECON 2005, Raleigh, India, 6–10 November 2005; pp. 79–83.
5. Hasanien, H.M.; Muyeen, S.M. Design optimization of controller parameters used in variable speed wind energy Conversion system by genetic algorithms. *IEEE Trans. Sustain. Energy* **2012**, *3*, 200–208. [CrossRef]
6. Ambia, M.N.; Hasanien, H.M.; Al-Durra, A.; Muyeen, S.M. Harmony search algorithm-based controller parameters optimization for a distributed-generation system. *IEEE Trans. Power Deliv.* **2015**, *30*, 246–255. [CrossRef]
7. Blasko, V.; Kaura, V. A novel control to actively damp resonance in input LC filter of a three-phase voltage source converter. *IEEE Trans. Ind. Appl.* **1997**, *33*, 542–550. [CrossRef]
8. Bingi, K.; Ibrahim, R.; Karsiti, M.N.; Chung, T.D.; Hassan, S.M. Optimal PID control of pH neutralization plant. In Proceedings of the 2nd IEEE International Symposium on Robotics and Manufacturing Automation, Ipoh, Malaysia, 25–27 September 2016; pp. 1–6.
9. Mishra, K.P.; Kumar, V.; Rana, K.P.S. Stiction combating intelligent controller tuning: A comparative study. In Proceedings of the International Conference on Advances in Computer Engineering and Application, Ghaziabad, India, 19–20 March 2015; pp. 534–541.
10. Ho, W.K.; Gan, O.P.; Tay, E.B.; Ang, E.L. Performance and gain and phase margins of well-known PID tuning formulas. *IEEE Trans. Control. Syst. Technol.* **1996**, *4*, 473–477. [CrossRef]
11. Costa, B.L.G.; Graciola, C.L.; Angélico, B.A.; Goedtel, A.; Castoldi, M.F. Metaheuristics Optimization Applied to PI controllers Tuning of a DTC-SVM Drive for Three-Phase Induction Motors. *Appl. Soft Comput.* **2017**, *62*, 776–788. [CrossRef]
12. Gizi, A.J.A.; Mustafa, M.; Jebur, H.H. A novel design of high-sensitive fuzzy pid controller. *Appl. Soft Comput.* **2014**, *24*, 794–805. [CrossRef]
13. Zemmit, A.; Messalti, S.; Harrag, A. A new improved dtc of doubly fed induction machine using ga based pi controller. *Ain Shams Eng. J.* **2017**, *4*, 1–9. [CrossRef]
14. Premkumar, K.; Manikandam, B. Speed control of brushless dc motor using bat algorithm optimized adaptive neuro-fuzzy inference system. *Appl. Soft Comput.* **2015**, *32*, 403–419. [CrossRef]
15. Mohamed, M.A.; Diab, A.A.; Rezk, H. Partial shading mitigation of PV systems via different meta-heuristic techniques. *Renew. Energy* **2019**, *130*, 1159–1175. [CrossRef]
16. Abdalla, O.; Rezk, H.; Ahmed, E.M. Wind driven optimization algorithm based global MPPT for PV system under non-uniform solar irradiance. *Sol. Energy* **2019**, *180*, 429–444. [CrossRef]
17. Oshaba, A.S.; Ali, E.S. Speed control of induction motor fed from wind turbine via particle swarm optimization based pi controller. *Res. J. Appl. Sci. Eng. Technol.* **2013**, *5*, 4594–4606. [CrossRef]
18. Douiri, M.R.; Cherkaoui, M. Neuro-genetic observer speed for direct torque neuro-fuzzy control of induction motor drive, Journal of circuits. *Syst. Comput.* **2012**, *21*, 1–18.
19. Ali, E.S. Speed control of induction motor supplied by wind turbine via imperialist competitive algorithm. *Energy* **2015**, *89*, 593–600. [CrossRef]
20. Rajasekar, N.; Sundaram, K.M. Feed-back controller design for variable voltage speed induction motor drive via ant colony optimization. *Appl. Soft Comput.* **2012**, *12*, 1566–1573. [CrossRef]
21. Sundareswaran, K.; Nayak, P.S. Ant Colony based feedback controller design for soft-starter fed induction motor drive. *Appl. Soft Comput.* **2012**, *12*, 1566–1573. [CrossRef]
22. Salvatore, N.; Caponio, A.; Neri, F.; Stasi, S.; Cascella, G.L. Optimization of delayed state kalman filter based algorithm via differential evolution for sensorless control of induction motors. *IEEE Trans. Ind. Electron.* **2010**, *57*, 385–394. [CrossRef]
23. Costa, B.L.G.; Angélico, B.A.; Goedtel, A.; Castoldi, M.F.; Graciola, C.L. Differential evolution applied to dtc drive for three-phase induction motors using adaptive state observer. *J. Control Autom. Electr. Syst.* **2015**, *26*, 403–420. [CrossRef]
24. Hato, M.M.; Bouallègue, S.; Ayadi, M. Water Cycle Algorithm-tuned PI Control pf a Doubly Fed Induction Generator for Wind Energy Conversion. In Proceedings of the 9th International Renewable Energy Congress (IREC), Hammamet, Tunisia, 20–22 March 2018; pp. 1–6.
25. Hato, M.M.; Bouallègue, S. Direct Power Control Optimization for Doubly Fed Induction Generator Based Wind Turbine Systems. *Math. Comput. Appl.* **2019**, *24*, 77.

26. Arfaoui, J.; Rezk, H.; Al-Dhaifallah, M.; Elyes, F.; Abdelkader, M. Numerical Performance Evaluation of Solar Photovoltaic Water Pumping System under Partial Shading Condition using Modern Optimization. *Mathematics* **2019**, *7*, 1123. [CrossRef]
27. Devanshu, A.; Singh, M.; Kumar, N. DSP based feedback linearization control of vector controlled induction motor drive. In Proceedings of the 2016 IEEE 1st International Conference on Power Electronics, Intelligent Control and Energy Systems (ICPEICES), Delhi, India, 4–6 July 2016; pp. 1–6.
28. Hamid, K.H.A.N. *Field Oriented Control, Application Note*; Polytechech: Clermont-Ferrand, France, 2008.
29. Devanshu, A.; Singh, M.; Kumar, N. Sliding Mode Control of Induction Motor Drive Based on Feedback Linearization. *IETE J. Res.* **2020**, *66*, 256–269. [CrossRef]
30. Kim, S.-H. Maximum torque control of an induction machine in the field weakening region. *IEEE Trans. Ind. Appl.* **1995**, *31*, 787–794.
31. Singh, B.; Shukla, S.; Chandra, A.; Al-Haddad, K. Loss minimization of two stage solar powered speed sensorless vector controlled induction motor drive. In Proceedings of the IECON 2016—42nd Annual Conference of the IEEE Industrial Electronics Society, Florence, Italy, 24–27 October 2016; pp. 1–6.
32. Salim, D.; Kheldoun, A.; Sadouni, R. Fuzzy indirect field oriented control of dual star induction motor water pumping system fed by photovoltaic generator. *Int. J. Eng. Intell. Syst. Electr. Eng. Commun.* **2015**, *23*, 63–76.
33. Singh, S.; Singh, B. Solar PV water pumping system with DC-Link voltage regulation. *Int. J. Power Electr.* **2015**, *7*, 72–85. [CrossRef]
34. Wanzeller, A.M. Current control loop for tracking of maximum power point supplied for photovoltaic array. *IEEE Trans. Instrum. Meas.* **2004**, *53*, 1304–1310. [CrossRef]
35. Marouani, R.; Mami, A. Voltage Oriented Control Applied to a Grid Connected Photovoltaic System with Maximum Power Point Tracking Technique. *Am. J. Appl. Sci.* **2010**, *7*, 1168–1173. [CrossRef]
36. Consoli, A.R. Experimental low-chattering sliding-mode control of a pm motor drive. *Eur. Power Electr.* **1991**, *1*, 13–18.
37. Fathy, A.; Al-Dhaifallah, M.; Rezk, H. Recent Coyote Algorithm-Based Energy Management Strategy for Enhancing Fuel Economy of Hybrid FC/Battery/SC System. *IEEE Access* **2019**, *7*, 179409–179419. [CrossRef]
38. Pierezan, J.; Coelho, L.D.S. Coyote optimization Algorithm: A new metaheuristic for global optimization problems. In Proceedings of the 2018 IEEE Congress on Evolutionary Computation (CEC), Rio de Janeiro, Brazil, 29 January 2018; pp. 1–8.

energies

Article

Performance Analysis of Axial-Flux Induction Motor with Skewed Rotor

Fatma Keskin Arabul *, Ibrahim Senol and Yasemin Oner

Department of Electrical Engineering, Yildiz Technical University, 34220 Istanbul, Turkey; senol@yildiz.edu.tr (I.S.); yoner@yildiz.edu.tr (Y.O.)
* Correspondence: fkeskin@yildiz.edu.tr

Received: 27 August 2020; Accepted: 21 September 2020; Published: 23 September 2020

Abstract: In recent years, with developing technology in the field of electrical machines, more efficient and high power density electric motors have been produced. The use of high energy efficiency motors gains importance due to the increase in global energy demand. The main purpose of this study was to design an Axial Flux Induction Motor (AFIM) with the same efficiency class as the Radial Flux Induction Motor (RFIM) in premium efficiency (IE3) class which is used commonly in industrial applications. Various AFIMs are designed with different rotor slot numbers and performance analyses as efficiency and torque ripple changes are investigated. It is known that torque ripple is one of the key parameters in electrical machine design which should be kept as low as possible without decreasing efficiency and torque. Accordingly, AFIMs' rotor slots are skewed considering the stator and rotor slot numbers. The use of a Soft Magnetic Composites (SMC) material in design is also investigated. As a result of the analyses, many premium efficiency classes for AFIMs are obtained. In addition, using SMC material and skewing the rotor slots provides that torque ripples be reduced.

Keywords: axial flux induction motor; finite element analysis; performance evaluation

1. Introduction

Today, more than 40% of the global energy consumption amount is consumed by induction motors and this rate exceeds 70% in the industry [1,2]. Additionally, these motors are key components of many industrial processes, with their reliability, and low cost of maintenance and construction [3,4]. Considering the amount of energy consumed by these motors, it is seen that even a small change in their efficiency will provide significant savings in worldwide energy consumption. The largest energy savings, particularly for medium and small power motors, arise for their higher efficiency classes [5,6]. Accordingly, the use of IE3 class motors has become mandatory due to the laws published in many countries. For instance, IE3 motors have been mandatory since 2011 in the United States and in Turkey, China, and the EU countries in 2015 [7–9]. Nowadays, energy consumption and environmental impacts are reduced with high efficiency motors. Additionally, motor reliability increases sustainable use and demand for investment [10]. In many countries around the world, many programs are encouraged to increase the use of high-efficiency motors. Among these programs, the Efficiency Increasing Project and the Efficient Motor Replacement programs are prominent in the world. Replacing existing low-efficiency motors with high-efficiency motors will result in significant energy savings even if same-sized motors are used [11,12]. Otherwise, if AFIMs are preferred over conventional RFIMs, more efficient and smaller volume motors can be designed [13,14].

With the developing technology, electric motors have a more compact structure and many studies are carried out to increase their efficiency. In many studies, instead of radial flux design of induction motors, axial flux design was found to have a more compact structure and it was concluded that their efficiency and torque density could be increased further [14,15].

AFIMs have the same operating principle as RFIM. However, design of these motors is quite different. The main difference is the magnetic flux direction. In conventional radial flux machines, the flux is in radial direction relative to the machine axis. The magnetic flux produced in AFIMs is in axial direction with respect to the machine axis.

In recent decades, AFIMs have been a popular research topic for researchers. Many studies have been conducted in the literature on AFIM design and control [16–18]. Among these studies, the most notable ones are the design and implementation of different structures. Additionally, many AFIMs have been designed for various applications such as pumps, electric vehicles, wind turbine, etc. [19,20] For instance, in electric vehicles, which are among the popular topics of today, AFIMs have different uses such as wheel-directly coupled, on-wheel, or main motor [21]. In some studies, different materials have been used, such as iron, magneto dielectric, superconductors, etc. [22,23].

In this study, AFIMs are designed to have the same efficiency class as RFIM in IE3 efficiency class used in industry. In addition, the rotor is skewed to minimize the torque ripple. In the literature, only stator slot numbers has been taken into consideration to determine the skew angles for the performance analysis of an induction motor [24–27]. In this study, and different from the literature, skew angles are selected considering both the ratio of stator and rotor slot numbers. Additionally, analyses are carried out using SMC material for stator, rotor, and both stator and rotor of the motor, which provides the best results in terms of efficiency and torque ripple. However, in the analysis results obtained, a considerable decrease in torque ripple over 11% is observed.

Following the introduction, Section 2 presents the topologies of AFMs and design parameters of AFIM. Section 3 introduces model properties, results, and discussion of all various types of AFIM models. Finally, the concluding remarks are presented in Section 4.

2. Methodology

In this section, considering the geometrical properties of the stator/rotor core, radial and axial flux machine equations are presented after the topologies of AFMs are introduced.

2.1. Axial Flux Machine Topologies

A machine with one air gap is the oldest and simplest structure of AFM, has a single-sided motor, and a single stator-single rotor (SSSR). The structure of this machine is easier due to a single air gap [28]. Generally, this machine is the best choice for low torque applications such as fans, pumps, food processors, etc. [29] Also, it can be said that single-sided AFIMs are more resistant to static eccentricity than conventional motors [30,31]. The disadvantage of this type of machine is that bearing life depends on their load. Active material utilization of the SSSR machine is higher [32]. In Figure 1, single-sided AFIM components are shown.

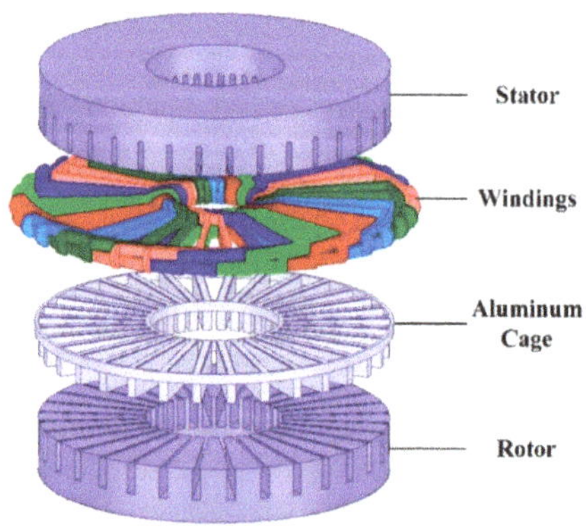

Figure 1. Single-sided Axial Flux Induction Motor (AFIM) components.

There are two air gaps in this type of machine. Both of these air gaps can be axially (double-sided motor), or one axially and the other radially. Such motors consist of a double stator-single rotor, with the rotor sandwiched between the stators, or single stator-double rotor with the stator sandwiched between the rotors. The advantages of double-sided motors include high torque density and balanced axial forces.

In this context, and in terms of economy, the production of two stators is more costly, especially in small powerful machines compared to the single-sided structure. However, the difference in production cost between single-sided motor and double-sided motor for high torque machines is decreasing. In double-sided motors, the moment of inertia is lower and the rotor is lighter [33].

Although the double-sided motor structure has better performance, for high-powered motors, the multi-air-gap disc structure is a better choice [34]. Multi air gap machines have two topologies that are determined by the number of stators and rotors. If the number of stators is more than the number of rotors, these machines are called external stator and internal rotor machines. If the number of rotors is more than the number of stators, these machines are called internal stator external rotor type machines. Internal stator external rotor type machines are preferred due to their high efficiency [35]. This topology can be defined as a concept rather than a machine type. The aim is to place the stators and rotors alternately to meet the requirements of the application. An advantage of this configuration is that it offers modularity [36].

In this study, single air-gap motor topology—which is also prominent in terms of ease of design and analysis—is chosen. It is an advantage that the volume of this structure is smaller. A conventional radial flux induction motor used industrially in the premium efficiency class is taken as a reference to the designed AFIMs.

2.2. Design Parameters of AFIM

The rotating magnetic field can be solved analytically by integrating the basic flux with respect to all radius (r) and pole form factor (α). The rotating magnetic field is within a pole range and electrical angle values are accepted. Axial rotating magnetic field is obtained as in Equation (1).

$$\phi_{ax} = \frac{2B_{\max}}{p}(r_2^2 - r_1^2)\cos\omega t \quad (1)$$

where ϕ_{ax} is axial rotating magnetic field in disk-like air gap; $B_{\max}$ is maximum induction in the air gap; p is number of poles; r_2 is outer radius of the core of axial induction motor and r_1 is inner radius of the core of axial induction motor.

Basically, the output power of electrical machines (P_2) is a function of flux per pole as in Equation (2).

$$P_2 = f(\phi) \quad (2)$$

Therefore, both types of induction motors can be compared using the flux equations of the rotating magnetic field. In 1986, Varga compared both types of motors using the flux equations of the rotating magnetic field in his study [37]. Thus, the flux equation of AFIM (Equation (1)) has been compared with the similar equation of RFIM (Equation (3)).

$$\phi_{rad} = \frac{4B_{\max}}{p}\mathrm{L}_r(r_r - r_o)\cos\omega t \quad (3)$$

where ϕ_{rad} is radial rotating magnetic field in cylindrical air gap; L_r is length of the radial induction motor, r_r is rotor radius of the radial induction motor; r_o is radius of shaft opening.

Flux values for two types of motors are equalized for predictive comparison ($\phi_{ax} = \phi_{rad}$), so Equation (4) becomes;

$$2\mathrm{L}_r(r_r - r_o) = r_2^2 - r_1^2 \quad (4)$$

In fact, both sides of equation 4 are cross-sectional areas for total magnetic flux in related type induction motors. In Figure 2, RFIM and AFIM geometries are shown. This comparison takes into account only the geometric properties of different types of induction motors under equal magnetic use.

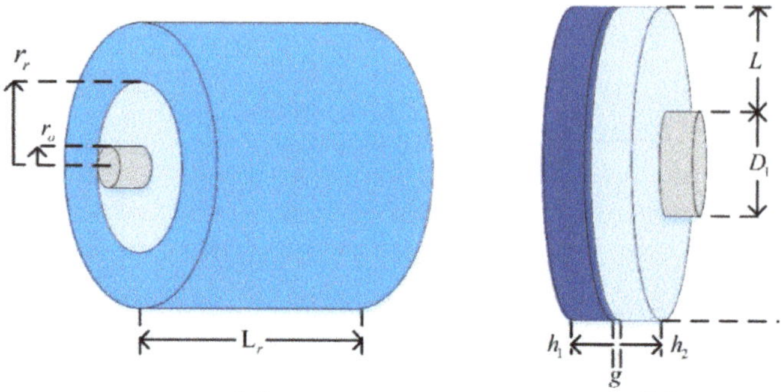

Figure 2. (**a**) Radial Flux Induction Motor (RFIM) geometry, (**b**) AFIM geometry.

The shaft opening increases the rotor radius r_r by r_o. The equations given AFIM have a single air gap and no volumetric assumptions are made. In Equation (5), apparent internal power is solved; r_1 and r_2 are halves of D_1 and D_2, respectively.

$$S_i = C_{ax}(D_2^2 - D_1^2)n_1 \tag{5}$$

where S_i is apparent internal power; C_{ax} is axial induction motor constant; D_2 is core outside diameter with no slots; D_1 is core inside diameter; n_1 is synchronous speed.

In Equation (6), C_{ax} is calculated as;

$$C_{ax} = \frac{\pi^2 \alpha k_w}{\sqrt{2}480} B_{\max} F_1 = \frac{B_{\max} F_1}{109.13} \tag{6}$$

where α is deflection, pole form factor; k_w is winding factor; and F_1 is distributed MMF; k_w and α values are taken 0.9 and 0.7, respectively, as the average values usually used [38].

In Equation (7), D_a refers to the average diameter of the core;

$$D_a = \frac{D_2 + D_1}{2} \tag{7}$$

In Equation (8), L refers to the radial width of the core;

$$L = \frac{D_2 - D_1}{2} \tag{8}$$

While establishing equations for diameter dimensions of AFIM, slots are neglected. The motor air gap is assumed to be constant when the slots are neglected. In these conditions, the outer diameter of the core D_2 can be found as in Equation (9).

$$D_2 = \sqrt{D_1^2 + \frac{2p\phi_{ax}}{B_{\max}}} \tag{9}$$

Core volume of the entire stator and rotor (V) is calculated in Equation (10).

$$V = V_s + V_r = \frac{\pi}{4}(D_2^2 - D_1^2)\sum_i h_i \tag{10}$$

where V_s is core volume of stator; V_r is core volume of rotor; $\sum_i h_i$ is total height of the machine.

Efficiency is calculated as in Equation (11) according to IEEE Standard 112-2017 [39].

$$\eta = \frac{P_{output}}{P_{input}} = \frac{P_{input} - \sum P_{losses}}{P_{input}} \tag{11}$$

where η is efficiency in percent; P_{output} is output power; P_{input} is inner output [39]. $\sum P_{losses}$ is total losses and presented as in Equation (12);

$$\sum P_{losses} = P_{core} + P_{cu} + P_{FW} + P_{STR} \tag{12}$$

where P_{core} is stator and rotor core losses; P_{cu} is stator and rotor copper losses; P_{FW} is mechanical losses as frictional and windage losses and P_{STR} is stray-load losses. The biggest cause of stray-load losses is harmonic energies when the motor is operating under load. If the load current consists of harmonic components, flux magnitude and waveform were distorted, which results in harmonic torques, vibrations and the noise [40,41]. In the IEEE Standard 112-2017, the assumed values for stray-load losses are shown in Table 1. According to the table, in the study, stray-load losses percent of rated load are taken as 1.8% for designed AFIMs [39].

Table 1. Assumed values for stray-load loss.

Machine Rating (kW)	Stray-Load Loss Percent of Rated Load
0.7457 to 90	1.8%
91 to 375	1.5%
376 to 1850	1.2%
1851 and greater	0.9%

In this study, transient analyses are carried out and torque values are calculated with the help of ANSYS Maxwell program. Torque ripple (τ_{ripple}) in percent is defined as the ratio of difference between maximum (τ_{max}) and minimum torque (τ_{min}) values to average torque (τ_{avg}) as in Equation (13).

$$\tau_{ripple} = \frac{\tau_{max} - \tau_{min}}{\tau_{avg}} \cdot 100 \tag{13}$$

AFIM results based on different rotor slot numbers are examined in the next section. In this study, the rotor slots are skewed to achieve better efficiency and fewer torque ripple. Skew angles are given for all slot numbers of the non-skew models, and analyses have been made by taking into account the stator and rotor slot numbers. Skew angle analyses adjusted according to stator slot number are given in Equations (14)–(16).

$$\beta_s = \frac{360}{N_s} \tag{14}$$

$$1.5 \cdot \beta_s = \frac{360}{N_s} = 15^{\circ} \tag{15}$$

$$2 \cdot \beta_s = \frac{360}{N_s} = 20^{\circ} \tag{16}$$

where β_s is skew angle according to stator slot number in degrees, N_s is stator slot number. By using Equations (14)–(16); 10°, 15° and 20° are selected. These values are the same for all design and analyzed for all of them.

Skew angle analyses adjusted according to rotor slot number are given in Equations (17) and (18).

$$\beta_r = \frac{360}{N_r} \tag{17}$$

$$2 \cdot \beta_r = \frac{360}{N_r} \tag{18}$$

where β_s is skew angle according to rotor slot number in degrees, N_r is rotor slot number.

In this context, geometries with three additional skew angles, β_s, $1.5 \cdot \beta_s$, $2 \cdot \beta_s$, β_r and $2 \cdot \beta_r$ are applied in the rotor slots of the 28, 30, 32, 34, 38, 40 and 42 slotted models, and analyses results are presented.

3. Results and Discussion

In this section, general design parameters of analyzed AFIMs with reference RFIM are presented. In addition, many AFIMs are designed to be in the same energy efficiency class as the RFIM referenced. While making these designs, no changes are made on the stator. When using the same stator, rotor slot numbers and skew angles are changed.

3.1. RFIM Model Details

This section provides information about the reference RFIM, which is widely used in the industrial applications. The analysis results of this motor, which is IE3 efficiency class with 93.2% efficiency and 7.49% torque ripple, was made by using the ANSYS Maxwell program. Analyses in which the motor model is suitable for 2D analysis are done in two dimensions. The ANSYS model of the model is shown in Figure 3. In addition, general design parameters of RFIM are shown in Table 2.

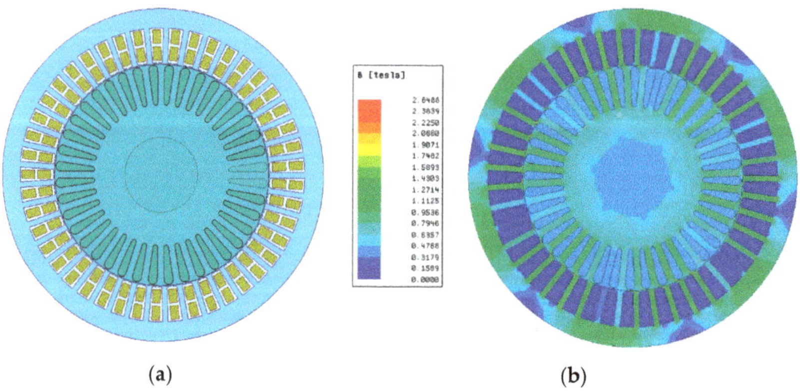

Figure 3. (**a**) RFIM model (**b**) Magnetic field density of RFIM.

Table 2. Design parameters of Radial Flux Induction Motor (RFIM).

Parameter	Value
Rated power	30 kW
Rated voltage	400 V
Rated speed	2980 rpm
Rated frequency	150 Hz
Stator core outer diameter	355 mm
Rotor core outer diameter	236.5 mm
Length of motor	250 mm
Stator/Rotor slot numbers	45/40
Stator material	Steel (M19_24G)

3.2. AFIM Model Details

In this section, Finite Element Analysis (FEA) is needed to perform electromagnetic analysis of AFIM. Unlike other types of machines, 3D analysis is mandatory because 2D analysis of such machines does not give detailed results. For this reason, analyses are performed in 3D Cartesian coordinate system and these analyses take a considerable amount of time due to high meshing. Also, for detailed analysis, the number of meshes is kept as high as the processor allowed. In this study, ANSYS Maxwell software is preferred to the FEA analysis, since it is suitable for magnetic field analysis of machines and has a wide usage area.

Firstly, an AFIM designed with the same number of stator and rotor slots. AFIM model and the magnetic field density of AFIM is shown in Figure 4. This model is designed with a non-skewed rotor. According to the analysis results, the efficiency of AFIM is 82.83%, while the torque ripple is 7.4%. Thereafter, in the next section, AFIMs are designed with different rotor slot numbers and various skewed rotors to increase the efficiency to IE3 efficiency class. Table 3 shows the general design parameters of the analyzed AFIMs.

Table 3. General design parameters of Axial Flux Induction Motor (AFIMs).

Parameter	Value
Rated power	30 kW
Rated voltage	400 V
Rated speed	2895 rpm
Rated frequency	150 Hz
Stator/Rotor core outer diameter	355 mm
Stator/Rotor core inner diameter	140 mm
Air-gap length	2 mm
Length of stator/rotor	60/55 mm
Stator slot numbers	36
Rotor slot numbers	28, 30, 32, 34, 38, 40, 42
Stator material	Steel (M19_24G)
Winding type	Whole Coiled
Number of conductors per slot	16
Number of strands	2
Conductor type	Copper

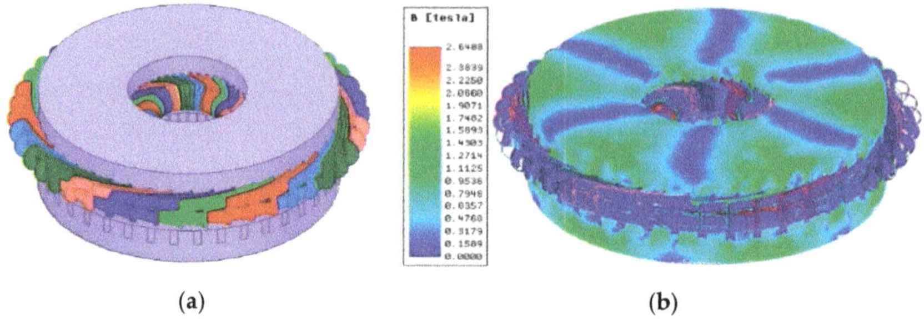

Figure 4. (**a**) AFIM model (**b**) Magnetic field density of AFIM.

3.3. *AFIM Analyze Results with Different Rotor Slot Numbers*

In this section, analyses are made according to seven different slot numbers for the rotor slots without skewing, 28, 30, 32, 34, 38, 40 and 42 slots, respectively. While determining these numbers, rotor slot widths are taken into account according to the machine size. Therefore, the maximum slot number is specified as 42 and the minimum as 28 for the number of rotor slots. Since the stator slot number is 36, 36-slot rotor design is not feasible, and because the stator slot and rotor slot areas are overlapped, the magnetic flux production is not possible.

It is known that torque ripple is one of the most challenging parameters in electrical machine designs. While making designs, decreasing this value is an aim. Therefore, torque ripples are taken into consideration while making comparisons. Figure 5 shows the efficiency and torque ripples of the analysis results according to 7 different slot numbers from 28 to 42. In Figure 5, torque ripple in 30 and 42 slotted AFIMs varies a lot compared to others. As the reason for this, it can be said that the number of these two slots is a multiple of 6; that is, the pole number. Thus, they are forced more in magnetic field transition. From the other five AFIMs, the torque ripples of only 28 and 34 slotted rotor designs are over 10%, while the other three designs' torque ripples at 9%. In this way, AFIM models with the best performance in terms of torque ripples are 32, 38 and 40 slotted models. There are six models in the same energy efficiency class (IE3) as RFIM. Only 34 slotted model is in a lower efficiency class (IE2).

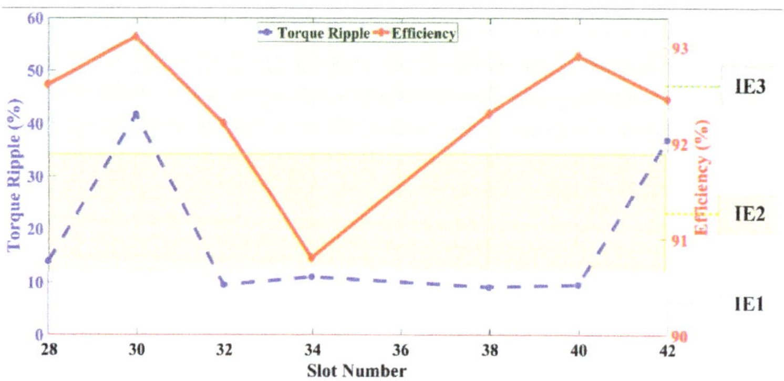

Figure 5. Efficiency and torque results of AFIM with different slot numbers.

As a result of the analyses made according to different rotor slot numbers, although it does not have the highest efficiency, it can be said that the 40 slotted model, which is in the same energy efficiency class and has little difference between its efficiency, is the best model among them since the torque ripple is much less.

3.4. Results of AFIM with Skewed Rotor

In the analysis of the rotor with 28 slot non-skew model, torque ripple 13.9%, efficiency 92.6% are obtained. The efficiency class of this model is in IE3 class. As the first analysis, skew angles are given to 28-slot AFIM. Accordingly, the skew angles in analyses are 10°, 12.86°, 15° and 20°. Torque ripple and efficiency results of the analyses are presented in Figure 6. Modeling of 25.72° skew is not feasible due to the overlay occurring between stator and rotor. In Figure 6, it is observed that there is no increase in efficiency when the rotor is skewed. However, when the models with skewed according to the number of stator slots are examined, it is seen that the efficiency decreases as the skew angle increases. On the other hand, when the skew is given according to the number of rotor slots, the efficiency keeps the same value.

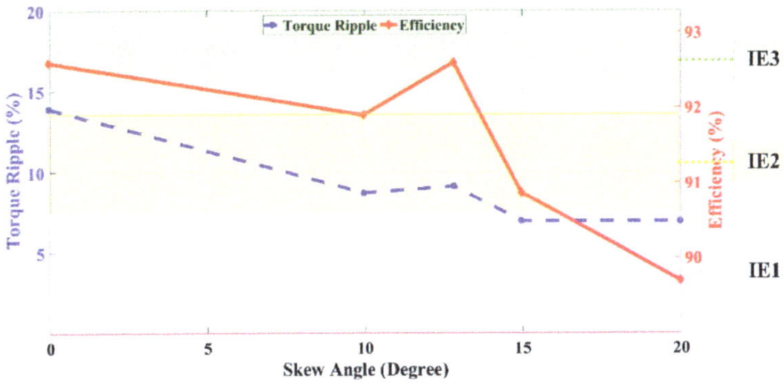

Figure 6. Efficiency and torque results of AFIM with 28 slots.

As a result of the non-skew model analyses, 30-slot AFIM model has the highest torque ripple value and the best efficiency. In the second analysis, this model is skewed. In addition to the non-skew model; in Figure 7 the torque ripple and the efficiency results of 10°, 12°, 15° and 20° skewed AFIM models are presented. In Figure 7, the torque ripple has decreased as the skew angle degree increases in analyses made only considering the number of stator slots. When analyses made according to the number of rotor slots are compared among themselves, it is seen that as the skew angle increases, the torque ripple decreases. In Figure 7, an increment in efficiency was not observed when the rotor is skewed. However, according to the skewed models, efficiency is increasing with the increment in the skew angle.

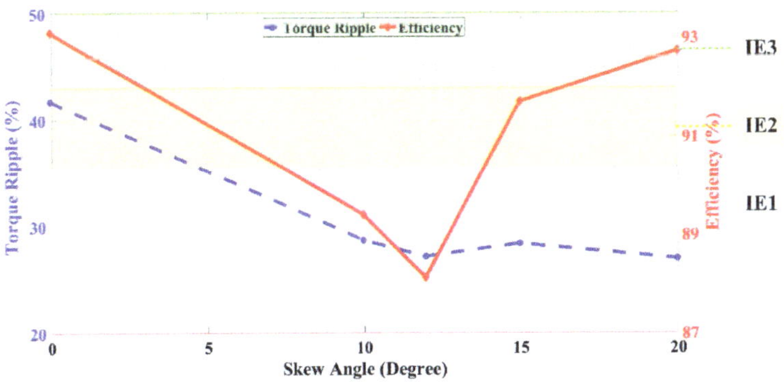

Figure 7. Efficiency and torque results of AFIM with 30 slots.

Considering the number of stators and rotor slots, the third analyses are done to 32-slot AFIM model with the skew angle of 10°, 11.25°, 15°, 20° and 22.5°. The non-skewed 32-slot AFIM model's torque ripple is 9.5% and the efficiency is 92.2%, which is in the IE3 class. Torque ripple and efficiency results of the analyses are shown in Figure 8. A comparison of the non-skew and 10° skew models shows that 10° skewed model has less torque ripple. On the other hand, a comparison of only skewed models shows that the torque ripple is lowest at 10° and highest at 15°. Comparing the 32-slot skewed models and the non-skew model, it is seen that the torque ripple decreases when the skew angle is 10°.

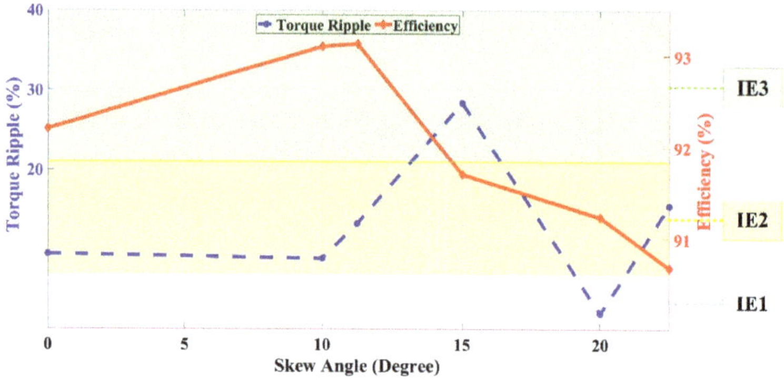

Figure 8. Efficiency and torque results of AFIM with 32 slots.

In addition, efficiency increases and the efficiency class increases from IE2 to IE3. Only from skewed models according to the number of stator slots, the efficiency decreases while the skew angle increases. On the other hand, according to the skewed models skewed with rotor slots number, the torque ripple increases and the efficiency decreases while the skew angle increases.

The fourth analysis is performed to 34-slot AFIM and the skew angles are calculated as 10°, 10.58°, 15°, 20° and 21.17°. In the 34-slot non-skewed model, the torque ripple is 11%, the efficiency is 90.8%, while the efficiency class is in the IE2 class. Torque ripple and efficiency results of the 32-slot AFIM analyses are shown in Figure 9. It is observed that the torque ripple does not fluctuate very much. Accordingly, it can be seen that the 15° skewed model is the most efficient model.

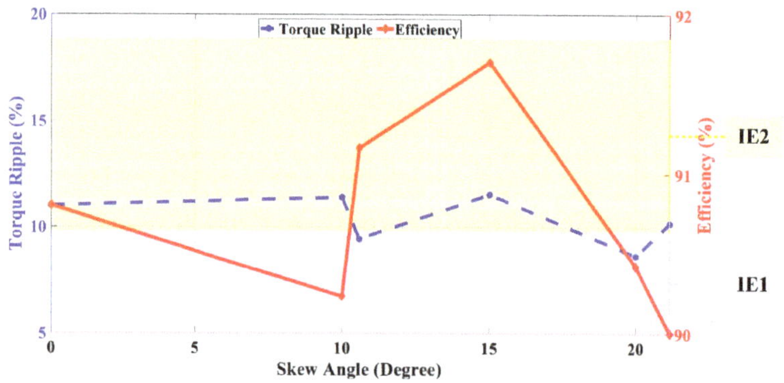

Figure 9. Efficiency and torque results of AFIM with 34 slots.

In the 5th analysis step, the 38-slot AFIM, which has the least torque ripple, is analyzed by considering the stator and rotor slots numbers. Accordingly, skew angles are determined as 9.47°, 10°, 15°, 18.95° and 20°. When the skewed analysis results are compared to the non-skewed model, the analysis with the lowest torque ripple is obtained when the skew angle is 9.47. Torque ripple and efficiency results of the 38-slot AFIM analyses are shown in Figure 10.

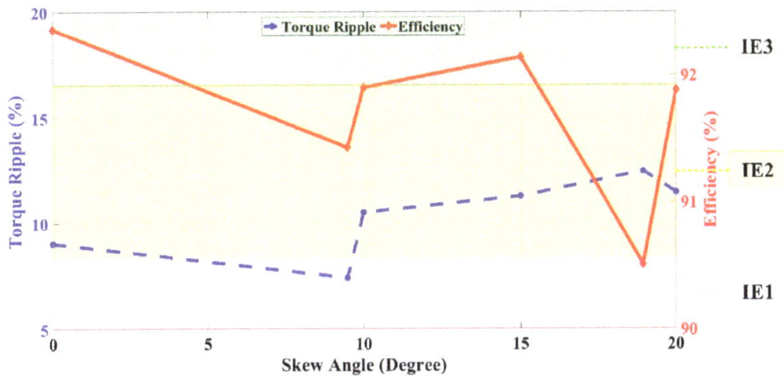

Figure 10. Efficiency and torque results of AFIM with 38 slots.

When the results of the skewed models of the 38-slot AFIM model according to the number of stator and rotor slots are examined, we do not see a positive effect on efficiency. Although the results are close to each other, the non-skewed model gives better results than skewed models.

In the sixth analysis step, the skew applied to the rotor of the 40-slot AFIM, which has the second best efficiency among non-skewed models. This model is also in IE3 energy efficiency class and the only model with low torque fluctuation. Skew angles for this model are calculated as 9°, 10°, 15°, 18° and 20°. Figure 11 shows the torque and efficiency results of the non-skewed and skewed models. In Figure 11, the 18° skewed model has the best torque ripple percentage. It can be seen that giving a skew to the rotor of 40-slot AFIM affects the efficiency negatively. However, the non-skewed, 9°, 10°, 15° and 18° skewed models are in IE3 energy efficiency class, other 10° and 20° skewed models are in IE2 energy efficiency class.

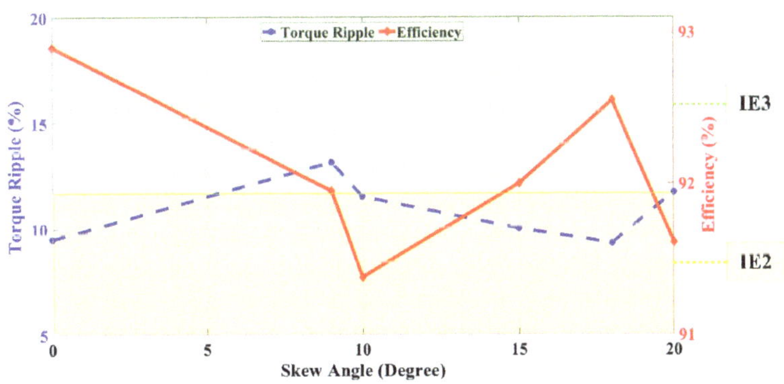

Figure 11. Efficiency and torque results of AFIM with 40 slots.

The last step of analyses among skewed models is done to 42-slot AFIM, which is another model with high torque ripple. Analyses results of non-skewed and skewed models of 42-slot AFIM are shown in Figure 12. The non-skewed model of 42-slot AFIM is in IE3 energy efficiency class. According to the analyses results, it is observed that there is a different situation between the efficiencies that are not encountered in the other slot numbers. Efficiency decreased in models with 10° and 20° skewed, whereas efficiency in skew angle 8.57° and 15° is significantly increased. In fact, it has been observed that three of these models are in the IE4 energy efficiency class. These are the highest efficient models achieved among analyses.

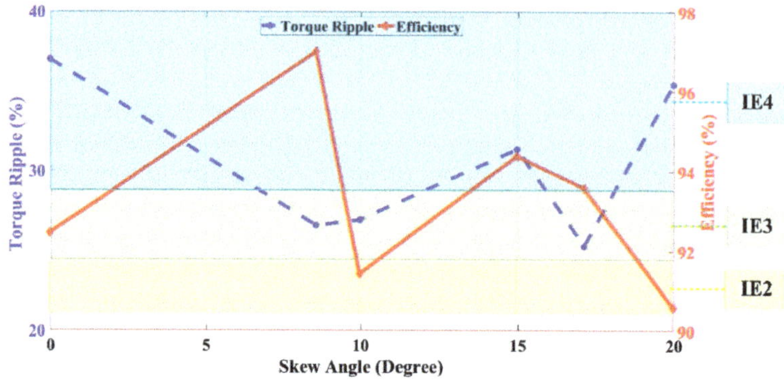

Figure 12. Efficiency and torque results of AFIM with 42 slots.

As a result of all analyses carried out, giving skews to the rotor of the motor affects the efficiency differently, and the torque ripple in different rotor slot numbers of AFIMs. In some AFIMs, skewing has a positive effect in terms of torque ripple and decreases the torque ripple, while in others it is observed that the torque ripple increases. In the same way, it has different effects in terms of efficiency. For example, the 8.57°, 15° and 17.14° skewed 42-slot AFIMs are given results in IE4 energy efficiency class. Also, the torque ripple of the same model decreases by 28% to 26.6%. Therefore, the best configuration among 42-slot models is obtained as 8.57° skew angle.

However, in the analyses performed both in terms of torque ripple and efficiency, the most optimum condition result is achieved in the design with a 10° skew given to the 32-slot AFIM. The torque ripple of this model is 8.93%, the efficiency is 93.1% and efficiency class of this motor is IE3. In the next section, the effect of the material on efficiency and torque ripple by using SMC material is examined.

3.5. Results of AFIM with SMC Material

In this section, the effect of the SMC material on efficiency and torque is examined. By this purpose, three analyses are done with using SMC material for stator, rotor and both of them. Somaloy 700-3P is selected as a SMC material with a density 7.57 g/cm^3 [42].

According to the analyses results, by changing the material, the efficiencies stayed in the same efficiency class as IE3. However, only SMC stator has the best efficiency value, both SMC stator&rotor has the best torque value. Compared to the non-skewed based model to both SMC stator&rotor used model over 11% of reduction is observed in torque ripple. Additionally, changing the material has a positive impact on the efficiency. Figure 13 shows the efficiency and torque results of SMC used AFIM by different parts.

In the X-axis of Figure 13, skew status and material of AFIM are defined as follows:

(a) Rotor is not skewed and both stator and rotor material are steel,
(b) Rotor is skewed and both stator and rotor material are steel,
(c) Rotor is skewed and stator material is SMC and rotor material is steel,
(d) Rotor is skewed and stator material is steel and rotor material is SMC,
(e) Rotor is skewed and both stator and rotor material are SMC.

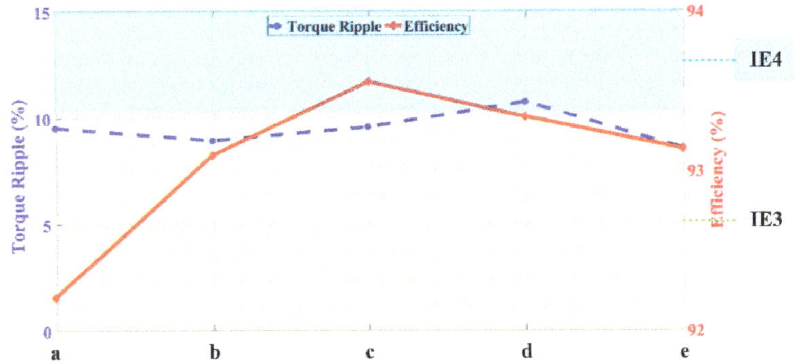

Figure 13. Efficiency and torque results of SMC used AFIM.

4. Conclusions

Since the aim in this study is to achieve an AFIM in the same energy efficiency class as RFIM, the efficiency of AFIMs with torque ripples are investigated. In this study, transient analyses are carried out and torque values are calculated with the help of ANSYS Maxwell program. Efficiency values of analysis results according to seven different slot numbers from 28 to 42 are presented. In addition to the efficiency values, the percentage torque ripples and energy efficiency class are also examined. To reduce the torque ripple values, rotor is skewed according to stator and rotor slot numbers. It was seen that the effects of skewing on efficiency and torque ripple are variable. Also, it was observed that to choose the skewing angle degrees, not only with stator slot number, but also rotor slot number could affect positive the performance of AFIMs. Using a SMC material on different parts of AFIM had a positive impact on performance analysis of a model, as reducing the torque ripple over 11% while the models are in premium efficiency class. For further studies, different materials could be used for various applications.

Author Contributions: F.K.A., I.S., Y.O. work on conceptualization, methodology, software, validation, formal analysis, writing—original draft preparation, writing—review and editing. All authors have read and agreed to the published version of the manuscript.

Funding: This research received no external funding.

Acknowledgments: The authors declare no acknowledgements.

Conflicts of Interest: The authors declare no conflict of interest.

References

1. De Almeida, A.T.; Ferreira, F.J.T.E.; Fong, J.A.C. Standards for Efficiency of Electric Motors. *IEEE Ind. Appl. Mag.* **2010**, *17*, 12–19. [CrossRef]
2. Ferreira, F.J.T.E.; Leprettre, B.; De Almeida, A.T. Comparison of Protection Requirements in IE2-, IE3-, and IE4-Class Motors. *IEEE Trans. Ind. Appl.* **2016**, *52*, 3603–3610. [CrossRef]
3. Elvira-Ortiz, D.A.; Morinigo-Sotelo, D.; Zorita-Lamadrid, A.L.; Osornio-Rios, R.A.; Romero-Troncoso, R.d.J. Fundamental frequency suppression for the detection of broken bar in induction motors at low slip and frequency. *Appl. Sci.* **2020**, *10*, 4160. [CrossRef]
4. Pietrowski, W.; Górny, K. Analysis of Torque Ripples of an Induction Motor Taking into Account a Inter-Turn Short-Circuit in a Stator Winding. *Energies* **2020**, *13*, 3626. [CrossRef]
5. Esen, G.K.; Ozdemir, E. A New Field Test Method for Determining Energy Efficiency of Induction Motor. *IEEE Trans. Instrum. Meas.* **2017**, *66*, 3170–3179. [CrossRef]
6. Parviainen, A. *Design of Axial-Flux Permanent-Magnet Low-Speed Machines and Performance Comparison between Radial-Flux and Axial-Flux Machines*; Lappenranta University of Technology: Lappeenranta, Finland, 2005; ISBN 9522140295. Available online: https://lutpub.lut.fi/bitstream/handle/10024/31185/TMP.objres.74.pdf?sequence=1 (accessed on 22 September 2020).

7. Boglietti, A.; Cavagnino, A.; Vaschetto, S. Induction motor EU standards for efficiency evaluation: The scenario after IEC 60034-2-1. In Proceedings of the IECON 2011—37th Annual Conference of the IEEE Industrial Electronics Society, Melbourne, VIC, Australia, 7–10 November 2011; pp. 2786–2791.
8. IEC-Governments & International Organizations, Examples by Industry Sector: Electric Motors-Measuring Efficiency. Available online: https://www.iec.ch/perspectives/government/sectors/electric_motors.htm (accessed on 18 March 2020).
9. Electric Motors | European Commission. Available online: https://ec.europa.eu/info/energy-climate-change-environment/standards-tools-and-labels/products-labelling-rules-and-requirements/energy-label-and-ecodesign/energy-efficient-products/electric-motors_en (accessed on 18 March 2020).
10. De Almeida, A.T.; Ferreira, F.J.T.E.; Baoming, G. Beyond induction motors—Technology trends to move up efficiency. *IEEE Trans. Ind. Appl.* **2014**, *50*, 2103–2114. [CrossRef]
11. Ferreira, F.J.T.E.; Baoming, G.; De Almeida, A.T. Reliability and operation of high-efficiency induction motors. In *Proceedings of the 2015 IEEE/IAS 51st Industrial and Commercial Power Systems Technical Conference, I and CPS 2015, Calgary, AB, Canada, 5–8 May 2015*; Institute of Electrical and Electronics Engineers Inc.: New York, NY, USA, 2015.
12. The Impact of New IE3 Requirements on Motors and Controls. Available online: https://www.totallyintegratedautomation.com/2017/11/impact-new-ie3-requirements-motors-controls/ (accessed on 18 March 2020).
13. Valtonen, M.; Parviainen, A.; Pyrhönen, J. Electromagnetic field analysis of 3D structure of axial-flux solid-rotor induction motor. In Proceedings of the International Symposium on Power Electronics, Electrical Drives, Automation and Motion, Taormina, Italy, 23–26 May 2006; SPEEDAM 2006. Volume 2006, pp. 174–178.
14. Evans, P.D.; Eastham, J.F. Disc-geometry homopolar synchronous machine. *IEE Proc. B Electr. Power Appl.* **1980**, *127*, 299–307. [CrossRef]
15. Valtonen, M.; Parviainen, A.; Pyrhönen, J. The effects of the number of rotor slots on the performance characteristics of axial-flux aluminium-cage solid-rotor core induction motor. In Proceedings of the IEEE International Electric Machines & Drives Conference, Antalya, Turkey, 3–5 May 2007; Volume 1, pp. 668–672.
16. Dianati, B.; Kahourzade, S.; Mahmoudi, A. Analytical design of axial-flux induction motors. In *Proceedings of the 2019 IEEE Vehicle Power and Propulsion Conference, VPPC 2019—Proceedings, Hanoi, Vietnam, 14–17 October 2019*; Institute of Electrical and Electronics Engineers Inc.: New York, NY, USA, 2019.
17. Rodger, D.; Coles, P.C.; Allen, N.; Lai, H.C.; Leonard, P.J.; Roberts, P. 3D finite element model of a disc induction machine. In Proceedings of the IEE Conference Publication, Cambridge, UK, 1–3 September 1997; pp. 148–149.
18. Valtonen, M.; Parviainen, A.; Pyrhönen, J. Inverter switching frequency effects on the rotor losses of an axial-flux solid-rotor core induction motor. In Proceedings of the POWERENG 2007—International Conference on Power Engineering—Energy and Electrical Drives Proceedings, Setubal, Portugal, 12–14 April 2007.
19. Dianati, B.; Kahourzade, S.; Mahmoudi, A. Optimization of Axial-Flux Induction Motors for the Application of Electric Vehicles Considering Driving Cycles. *IEEE Trans. Energy Convers.* **2020**, *35*, 1522–1533. [CrossRef]
20. Mei, J.; Lee, C.H.T.; Kirtley, J.L. Design of axial flux induction motor with reduced back iron for electric vehicles. *IEEE Trans. Veh. Technol.* **2020**, *69*, 293–301. [CrossRef]
21. Benoudjit, A.; Guettafi, A.; Nait Saïd, N. Axial flux induction motor for on-wheel drive propulsion system. *Electr. Mach. Power Syst.* **2000**, *28*, 1107–1125. [CrossRef]
22. Álvarez, A.; Suárez, P.; Cáceres, D.; Cordero, E.; Ceballos, J.M.; Pérez, B. Disk-shaped superconducting rotor under a rotating magnetic field: Speed dependence. *IEEE Trans. Appl. Supercond.* **2005**, *15*, 2174–2177. [CrossRef]
23. Kubzdela, S.; Weglinski, B. Magnetodielectrics in Induction Motors with Disk Rotor. *IEEE Trans. Magn.* **1987**, *24*, 635–638. [CrossRef]
24. Beleiu, H.G.; Maier, V.; Pavel, S.G.; Birou, I.; Pica, C.S.; Darab, P.C. Harmonics consequences on drive systems with induction motor. *Appl. Sci.* **2020**, *10*, 1528. [CrossRef]
25. Carbonieri, M.; Bianchi, N. Induction motor rotor losses analysis methods using finite element. In Proceedings of the 2020 IEEE International Conference on Industrial Technology (ICIT), Buenos Aires, Argentina, 26–28 February 2020; pp. 187–192.

26. Nobahari, A.; Darabi, A.; Hassannia, A. Various skewing arrangements and relative position of dual rotor of an axial flux induction motor, modelling and performance evaluation. *IET Electr. Power Appl.* **2018**, *12*, 575–580. [CrossRef]
27. Le Besnerais, J.; Lanfranchi, V.; Hecquet, M.; Romary, R.; Brochet, P. Optimal slot opening width for magnetic noise reduction in induction motors. *IEEE Trans. Magn.* **2009**, *45*, 3131–3136. [CrossRef]
28. Hecker, Q.; Igelspacher, J.; Herzog, H.G. Parameter identification of an axial-flux induction machine by winding functions. In Proceedings of the 19th International Conference on Electrical Machines- ICEM 2010, Rome, Italy, 6–8 September 2010.
29. Chan, C.C. Axial-field electrical machines—Design and applications. *IEEE Trans. Energy Convers.* **1987**. [CrossRef]
30. Nasiri-Gheidari, Z.; Lesani, H. New design solution for static eccentricity in single stator-single rotor axial flux induction motors. *IET Electr. Power Appl.* **2013**, *7*, 523–534. [CrossRef]
31. Nobahari, A.; Darabi, A.; Hassannia, A. Axial flux induction motor, design and evaluation of steady state modeling using equivalent circuit. In Proceedings of the 8th Power Electronics, Drive Systems and Technologies Conference, PEDSTC 2017, Mashhad, Iran, 14–16 February 2017.
32. Patterson, D.J.; Colton, J.L.; Mularcik, B.; Kennedy, B.J.; Camilleri, S.; Rohoza, R. A comparison of radial and axial flux structures in electrical machines. In Proceedings of the 2009 IEEE International Electric Machines and Drives Conference, IEMDC '09, Miami, FL, USA, 3–6 May 2009; pp. 1029–1035.
33. Mirzaei, M.; Mirsalim, M.; Abdollahi, S.E. Analytical modeling of axial air gap solid rotor induction machines using a quasi-three-dimensional method. *IEEE Trans. Magn.* **2007**, *43*, 3237–3242. [CrossRef]
34. Nasiri-Gheidari, Z.; Lesani, H. A survey on axial flux induction motors. *Prz. Elektrotechniczny* **2012**, *2*, 300–305.
35. Babu, V.R.; Soni, M.P. Modelling of Twin Rotor Axial Flux Induction Machine and its Application as Differential in Electrical Vehicles. *IJIREEICE* **2017**, *5*, 118–127. [CrossRef]
36. Egea, A. Overview of Axial Flux Machines. *Electr. Energy Mag.* **2013**, *4*, 2–13.
37. Varga, J.S. Magnetic and Dimensional Properties of Axial Induction Motors. *IEEE Trans. Energy Convers.* **1986**, *EC-1*, 137–144. [CrossRef]
38. Pyrhönen, J.; Jokinen, T.; Hrabovcová, V. *Design of Rotating Electrical Machines*; John Wiley & Sons: West Sussex, UK, 2008; ISBN 9780470695166.
39. IEEE. *IEEE Standard Test Procedure for Polyphase Induction Motors and Generators*; IEEE Std 112-2017 (Revision of IEEE Std 112-2004); IEEE: Piscataway, NJ, USA, 2018; pp. 1–115. [CrossRef]
40. Yamazaki, K.; Haruishi, Y. Stray load loss analysis of induction motor comparison of measurement due to IEEE standard 112 and direct calculation by finite element method. In Proceedings of the IEEE International Electric Machines and Drives Conference, IEMDC'03, Madison, WI, USA, 1–4 June 2003; Volume 1, pp. 285–290.
41. Levi, E.; Lamine, A.; Cavagnino, A. Impact of stray load losses on vector control accuracy in current-fed induction motor drives. *IEEE Trans. Energy Convers.* **2006**, *21*, 442–450. [CrossRef]
42. Somaloy® Quick Guide. Available online: https://www.hoganas.com/globalassets/download-media/sharepoint/brochures-and-datasheets---all-documents/somaloy-quick-guide_july_2015_1122hog.pdf (accessed on 20 September 2020).

energies

MDPI

Article

Design and Performance Assessment of a Small-Scale Ferrite-PM Flux Reversal Wind Generator

Bharathi Manne [1,*], Malligunta Kiran Kumar [1] and Udochukwu B. Akuru [2,3,*]

1 Electrical and Electronic Engineering, Koneru Lakshmaiah Educational Foundation, Guntur 522502, India; mkkumar@kluniversity.in
2 Department of Electrical Engineering, Tshwane University of Technology, Pretoria 0183, South Africa
3 Department of Electrical Engineering, University of Nigeria, Nsukka, Enugu State 410001, Nigeria
* Correspondence: bharathi.manne1994@gmail.com (B.M.); akuruub@tut.ac.za (U.B.A.); Tel.: +91-70-755-50797 (B.M.)

Received: 4 September 2020; Accepted: 29 September 2020; Published: 23 October 2020

Abstract: Currently, there is increasing research interest in harnessing wind energy for power generation by means of non-conventional electrical machines e.g., flux-reversal machines. The flux reversal machine is usually designed using scarce rare–earth permanent magnet material which may be unattractive in terms of machine cost. In this study, an attempt is made to re-design the flux reversal machine with non-rare-earth ferrite permanent magnet for wind energy applications. Because these machines possess high cogging torque, which results in vibration and noise, that are detrimental to the machine performance, especially at low speeds, a novel combined skewed and circumferential rotor pole pairing method is developed. The proposed cogging torque reduction method is implemented in 2-dimensional finite element analysis modeling and comparatively analyzed with other existing stand-alone methods viz., skewing, and rotor pole pairing. The results show that the proposed method led to 94.8% and 71% reduction in the cogging torque and torque ripple compared to the reference generator, respectively. However, the calculated torque density is reduced by 13%. Overall, the electromagnetic performance of the proposed ferrite PM machine exhibits desirable qualities as an alternative design for the direct drive wind generator.

Keywords: cogging torque; ferrite PM; finite element analysis; flux reversal machine; non-rare earth; wind energy

---

1. Introduction

The design of permanent magnet (PM) machines keeps growing significantly by the day due to advantages of high-power density, high-efficiency, and torque density, useful for variable speed drives in wind energy applications. The need for increased use of wind energy generation is well established, due to its availability throughout the day and the possibility for large megawatt generation. Globally, installed wind power capacity has grown exponentially in recent years as shown in Figure 1, captured from the recently released Renewables 2020 Global Status Report renewables figure [1]. Wind power generation cost and reliability are critical factors that have to be considered in wind generator designs, and as such attention is usually focused in the direction of the existing drive concepts [1]. The various wind generator drive concepts are high-speed (HS), medium-speed (MS), and low-speed (LS) as characterized in Table 1 [2].

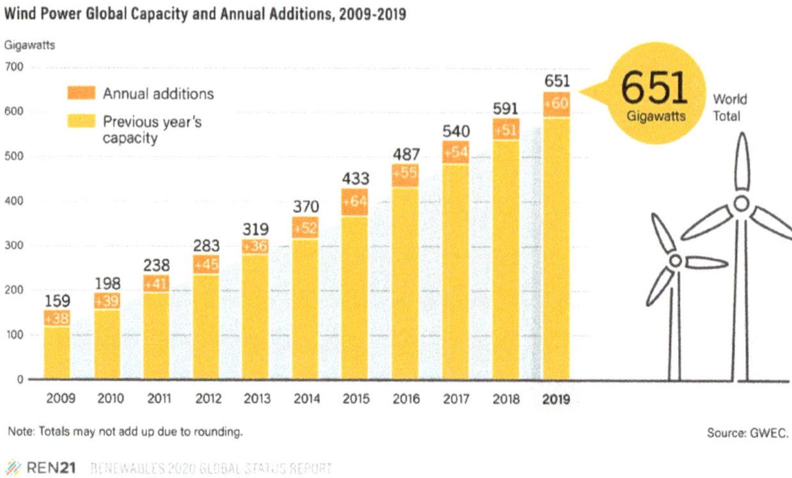

Figure 1. Global total installed capacity [1].

Table 1. Characteristics of wind turbine drive-trains [2].

Parameter	HS	MS	LS
Speed margin	600–2000 r/min	40–600 r/min	4–35 r/min *
Mass	Lightest	Intermediate	Heaviest
Size	Smallest	Intermediate	Largest
Gearbox presence	Yes (3G ¤)	Yes (1G/2G)	Absent
Generator type	IG/SG §	IG/SG	SG
Mechanical losses	High	Intermediate	Lowest
Electrical losses	Lowest	Intermediate	Highest
Cost	Gearbox #	Intermediate	Generator #

* Depends on operating power level. ¤ G represents the stage of the gearbox. § IG = induction generator, SG = synchronous generator. # highest cost.

PMs are classified into rare-earth (RE) and non-rare-earth (NRE) materials. In recent years, the unaffordability of RE PM prices forced the electrical machine designers to focus on inexpensive NRE materials like ferrite PMs and these materials are more desirable for the industry [3]. The trending flux-reversal machine (FRM), which is a pre-dominantly PM type machine, has been reported in some studies for rare-earth PM excited wind energy applications [4,5] and direct-drive drive-trains [6,7]. A direct-driven wind turbine generator is important since it generates electricity through the wind turbine drivetrain by eliminating the gearbox. The advantages of eliminating mechanical gearboxes from the wind generator system include a reduction in the installation costs due to a lesser number of components, lower energy costs due to a reduction in losses, and reduction in maintenance costs due to the simplistic design [6].

Depending on the variation in the rated speed of the wind turbine, direct drive is in the range of 120–500 r/min, with a corresponding generator power range of 0.5–3 KW [6]. The conventional synchronous and induction generators are not suitable for LS and MS applications [7]. The majority of LS and MS applications are handled by mechanical drives (mechanical gear and high-speed motor). For LS, existing generators require a large number of stator slots and poles, resulting in a machine with a large air-gap diameter [7].

Meanwhile, ferrite PM materials offer less expensive PM options for machine design but are more demagnetization prone due to the low coercivity compared to RE PM materials [8]. The affordability of NRE materials is mainly due to its availability as shown in Figure 2 [9].

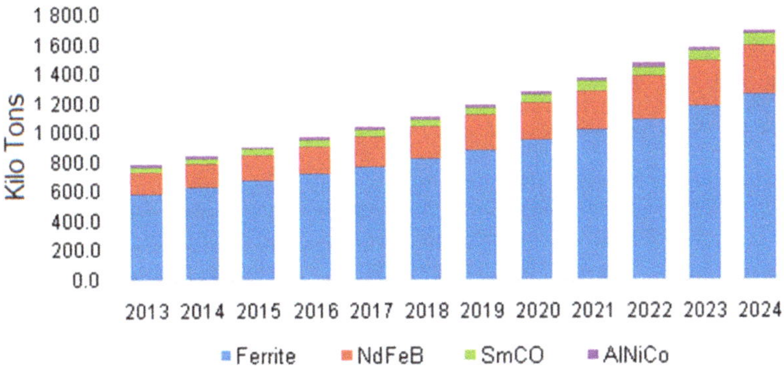

Figure 2. Global permanent magnet (PM) materials market capacity, 2013–2024 [9].

The FRM machine belongs to the category of double salient permanent magnet (DSPM) machines with PMs on the stator pole surface. Most of the PM machines employ interior rotor PM or surface mounted rotor PMs. Here, in this PMs are in the rotating part, which may cause demagnetization problems, limitations on mechanical instability, and poor thermal dissipation. These structures are not favorable for HS and MS applications [10,11]. To this end, stator-mounted permanent magnet (SMPM) machines were introduced, they are basically, flux switching machines (FSM) and flux reversal machines (FRM). These machines generally have similar features such as easy rotor structure, short winding terminals, good at thermal conditions [12]. FSM as a single-phase flux switch alternator and was introduced by Rauch and Johnson in 1955. It exhibited poor rotor volume usage causing stator vibrations and difficulty in stator fabrication [13]. To overcome these difficulties in the manufacturing process and to improve the torque (power) density, Deodhar et al. introduced single-phase FRM in 1997 for automobile applications to replace standard claw pole alternator [14,15].

The basic 3-phase FRM of 6/8 slot-pole combination with concentrated winding was introduced by Wang et al. [16,17]. In this machine, the design was optimized from single phase to 3-phase to ensure maximum PM flux-linkages in the stator winding, lighter PM, and low cogging torque (CT). However, this type of PM machines exhibit high torque ripple and lead to vibration and noise. The main cause of the torque ripple is the cogging torque (CT). The CT is mainly position-dependent and load-independent torque, caused due to the PMs, it mainly deteriorates the machine performance especially at LS and MS. More and more researchers dedicated themselves to suppress the CT in RE FRM while maintaining all other machine performances [18–22]. As seen from the literature, the CT reduction in the FRM plays major role in-terms of ripple-torque. Most of the CT minimization methods for FRM, published in the literature are some auxiliary techniques like notching, pole pairing, and skewing [18–21]. These techniques may be limited for the FRM as nothing serious has been reported for NRE variants. Besides, FRM requires less number of stator slots and a large number of rotor poles, which qualifies them for direct-driven wind energy generators [5,7].

The main contribution of this paper is to reduce the cogging torque and torque ripple by a combination of techniques in the proposed ferrite PM FRM while maintaining respectable electromagnetic performance compared to the basic machine as necessary for direct-drive wind generator applications. Thus, an overview of the FRM technology including working principle, machine capabilities using a flux-mmf diagram (FMDT), as well as design features based on different RE topologies, are discussed in Section 2. In Section 3, the finite element analyses (FEA) on two existing stand-alone cogging torque minimization methods, as well as the novel method combining these two methods are undertaken. The power generating performance of all the four generators considered in this paper in terms of output voltage, power density, voltage regulation, efficiency at the rated condition, and overload/speed capabilities are discussed and compared by FEA. In Section 4, some concluding remarks are given.

2. Overview of RE-PM FRMs

2.1. Structure and Configuration of RE-PM FRM

An early example of 6/4 pole double salient non-rotating PM type motor whose PMs are installed in the stator yoke is illustrated in Figure 3. This double salient structure enables the superior performance of torque production, small frame sizes and qualifies the motor as a potential alternative to existing servo drives, variable speed drives, as well as for satisfying the increasing demand in future automobiles. The experimental test results have been encouraging, demonstrating twice the output capability, with higher efficiency and power density when compared with the induction motor [11]. A new brushless double-salient 2/3 pole FRM, as depicted in Figure 4, is designed, analyzed, and fabricated based on a single-phase prototype machine for high-speed generator. FMDT is employed to analyze the qualitative performance of FRM with other types of brushless machines [15].

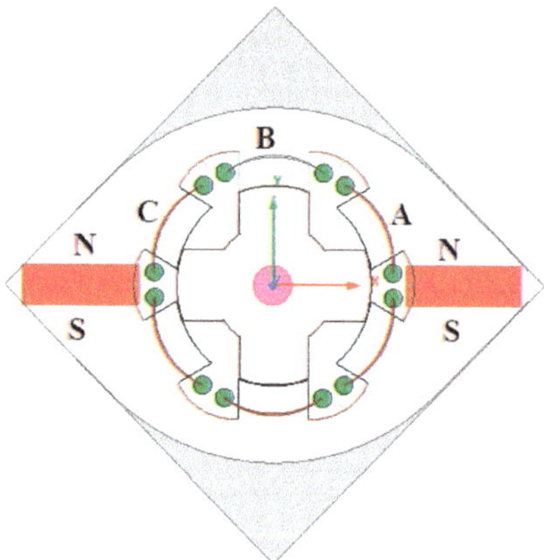

Figure 3. 3-phase, six/four-pole double salient permanent magnet (DSPM) machine [11].

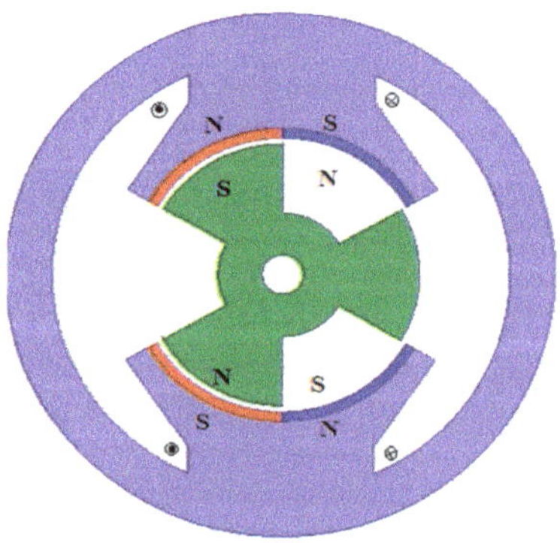

Figure 4. 1-phase, two/three-pole stator-PM generator [15].

Three-phase 6/8 FRM has been introduced by Wang as depicted in Figure 5, where the magnetic field distribution, self and mutual inductances, cogging torque variations with rotor positions are analyzed through 2D FEA [16,17]. In addition, 2-phase, 3-phase, and 5-phase pole combinations with suitable rotor skewing techniques are documented to minimize the cogging torque. The machine capabilities in terms of low rotor inertia, low electrical time constant, and high torque density, as proven by FEA, had been discussed.

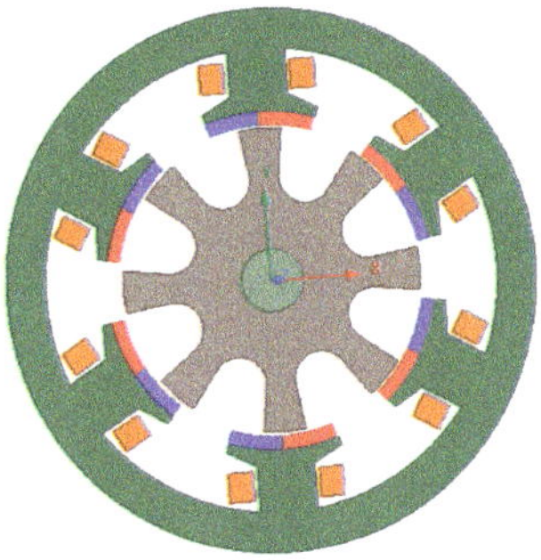

Figure 5. 3-phase, eight/six-pole flux-reversal machine (FRM) [16,17].

Another example of FRM for the servo drive application is depicted in Figure 6 with 12/28 poles [23]. The design specifications and operation of FRM for low-speed, high-torque applications are explained in [23]. The inset type PM structure is proposed to reduce the flux fringing of 12/40 pole FRM for servo drives (LS). Through FEA, it is shown that this configuration has achieved high efficiency, high torque density, and less than 3% torque ripple with three-phase sinusoidal vector control. Already, the candidature of the FRM for wind applications is growing as shown by some studies [24,25]. A 3-phase, 6/14 pole, 2.4 kW, 214 r/min outer rotor FRM has been introduced by D. S. More [24]. Through experimentation, it is concluded that inner rotor FRM has 1.25 times lower power density than outer rotor FRM. These two types of machines are depicted in Figures 7 and 8. Power density comparisons have been made in [25]. Compared with other types of DSPM, it is found that FRM has higher power density [5] for fractional-slot concentrated winding (FSCW) of FRM and permanent magnet synchronous machine (PMSM). Through experimentation, it is found that FRM has 1.5 times higher power density. Both the machine efficiencies are approximately the same [25]. To further improve the power density of FRM, a full-pitched winding (FPW) is incorporated in the stator of the FRM as shown in Figure 9 [26].

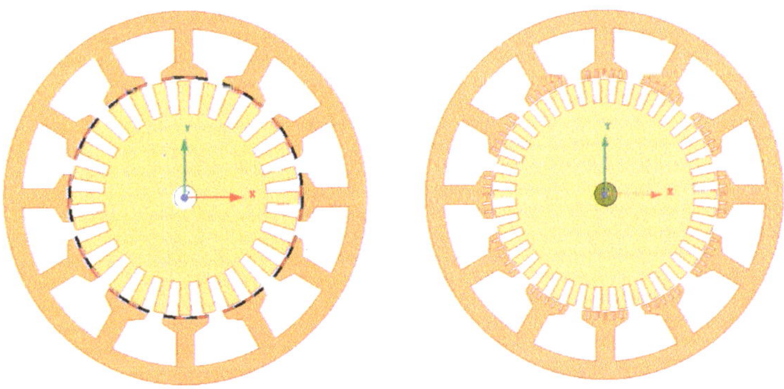

(a) (b)

Figure 6. Various pole and PM arrangements of the FRM machines [23]; (a) PMs on stator pole; (b) inset PMs on stator.

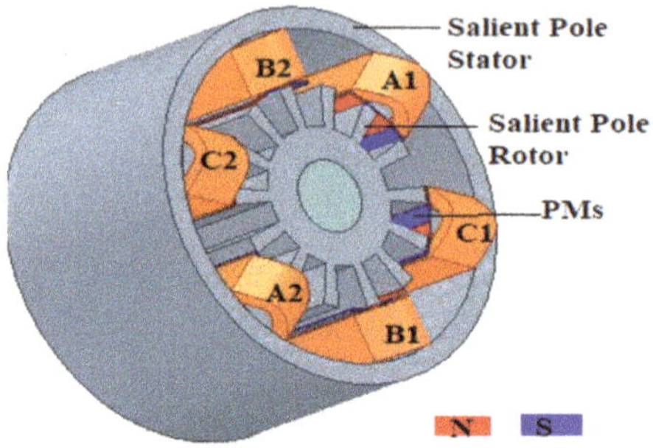

Figure 7. A typical 3D Structural view of 6/14 inner pole rotor of FRM [24].

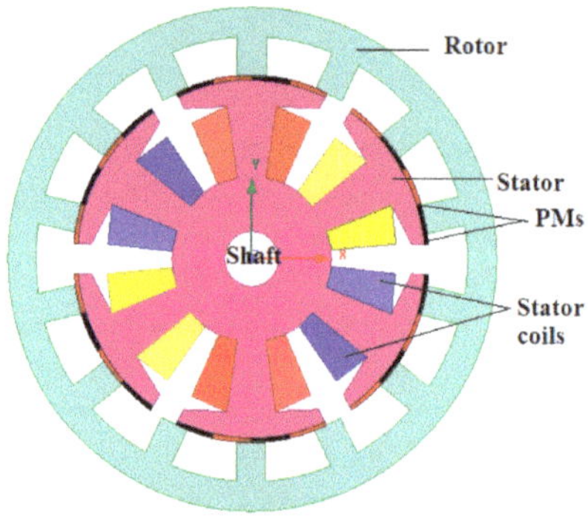

Figure 8. Cross-section of 3phase, 6/14 outer pole rotor FRM [24].

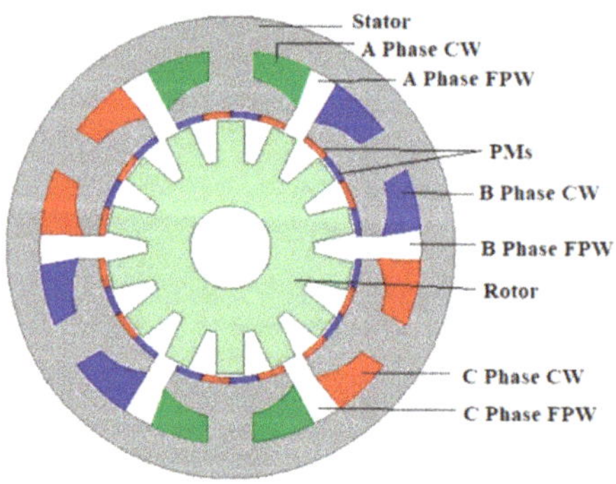

Figure 9. Structural-view of 6/14 FRM [25,26]; FPW: Full-pitched winding, CW: Concentrated winding.

Comparing different topologies of the FRM, that is, the consequent-pole permanent magnet (CPM) topology with the SMPM topology, as illustrated in Figure 10, it was concluded that CPM topology improves the torque performance and reduces the magnetic volume [27]. To reduce the cost of PMs, consequent-pole transverse-flux permanent magnet linear machine (TFPMLM) is proposed, partially replacing PM Poles by soft magnetic iron in [27,28].

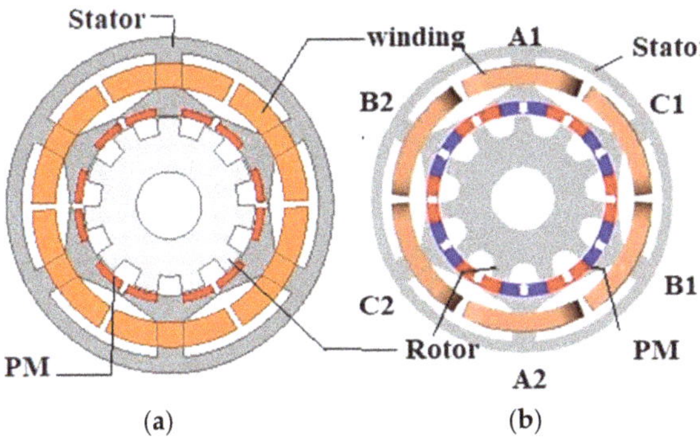

Figure 10. Typical topologies of FRM: (**a**) consequent-pole permanent magnet (CPM)-FRM; (**b**) stator-mounted permanent magnet (SMPM)-FRM [27,28].

Introducing a small space-gap between the adjacent PMs belonging to the same stator pole shoe, as illustrated in Figure 11, has helped to improve the phase back-emf and cogging torque [29]. FRM designed with soft magnetic composite materials for fans exhibited a significant increase in efficiency while producing high power density and reducing the usage of PMs compared to the existing PM type machines [30]. Based on electromagnetic compatibility, high-speed FRM topology has been proposed for the angular grinder as shown in Figure 12 and experimental results show that the efficiency of the developed machine is higher than that produced in induction and brushless type motors [31].

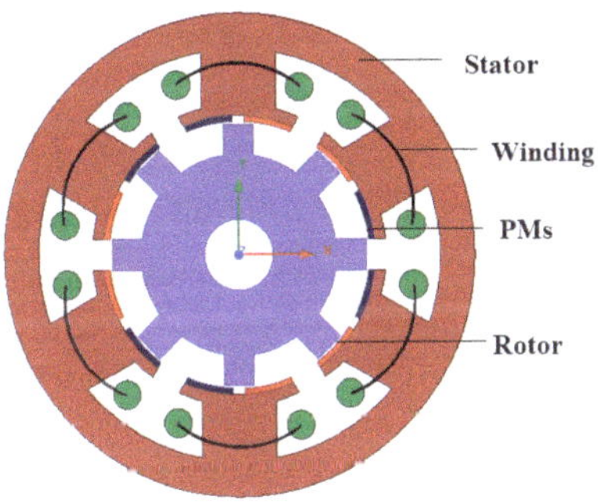

Figure 11. 3-phase, 6-stator-pole/8-rotor-pole FRM [29].

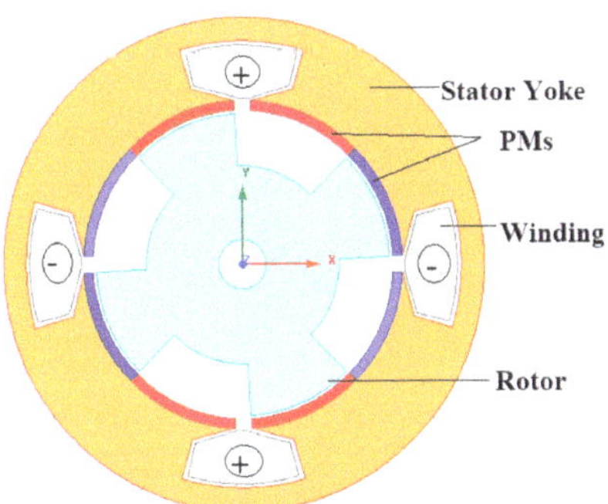

Figure 12. Cross-section of 1-phase, FRM motor [31].

Comparisons with the different configurations of the FRM showed that the stator flux linkages of the 6/14 FRM is doubled compared to that of the 12/16 pole FRM as illustrated in Figure 13 [32]. This is because in the 12/16 pole FRM, the stator teeth surface occupies around 2/3 of the inner stator surface and 1/3 of the stator inner surface is wasted. Accordingly, it causes high cogging torque (CT), vibrations, and acoustic noise. To overcome all the constructional issues of the 12/16 pole FRM, Dmitrievskii introduced the inner stator surface 12/10 pole FRM in 2018 as shown in Figure 14 [32]. Acoustic noise and CT are further reduced by this configuration for wind applications.

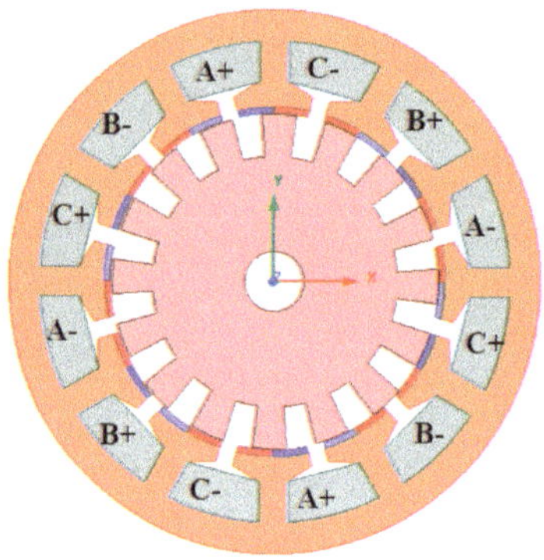

Figure 13. Schematic representation of 12-stator-slot/16-pole-rotor of FRM [32].

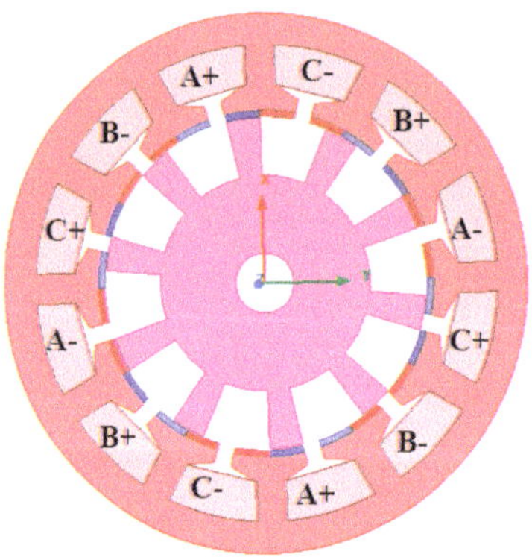

Figure 14. Configuration of a 12-stator-slot/10-pole-rotor of FRM [32].

A 3 kW, 200 r/min, 48/46 pole FRM has been projected for direct-drive wind energy [32] and is depicted in Figure 15, FEA calculations are done for FRM and PMSM with the same machine dimensions, power ratings, and speed. It is concluded that core volume is reduced by 25.6% and PMs required for FRM is five times less than that for PMSM and correspondingly, the cost of the active material like the PMs becomes twice as low as for PMSM. Then, the number of poles of FRM is higher, the frequency of FRM is thrice as high at the same speed of both the machines, and hence FRM is best suitable for variable speed applications than PMSM. The FRM cooling is simpler, and it can run at higher ambient temperatures. The FRM efficiency is 2.3% higher than the PMSM for gearless wind energy applications [31]. More [5,6] and Pellegrino [7] introduced fictitious electrical gear of 6/14 FRM for direct drivetrains i.e., low-speed high torque applications. It is concluded with fabrication results that FRM can be considered as PMSM with an inbuilt gearing effect [5,6]. The various design parameters influencing FRM performance are also analyzed [7]. A new proposed FRM structure is examined to improve the CT profile by changing the PM thickness and rotor side geometrical parameters for wind generators [33].

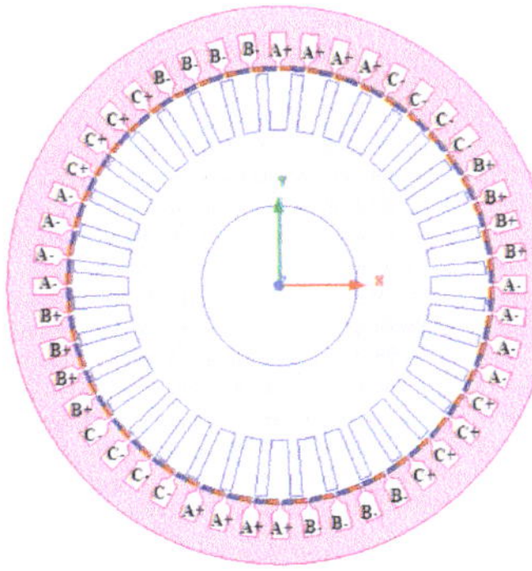

Figure 15. 3-phase, 48/46-pole FRM generator for the direct drive [32].

2.2. Basic Principle, Design Topologies, and Performance of FRM

FRMs are hybrid machines combining the advantages of a switched reluctance machine (SRM) and DSPM into one machine. Both the rotor and stator poles have double salient structures. For this non-conventional type of machine, the operational capabilities are analyzed by the flux-mmf diagram technique (FMDT) and implemented using FEA by Ion Boldea in 1996 [14,15]. FMDT has its origin in the ψ–I diagram and can predict the periodic variation of phase mmf and flux variations over an electrical cycle. The FMDT for three double salient type machines for comparisons is illustrated in Figure 16. The area enclosed by the flux-mmf loop is the average energy converted over an electrical period and it specifies the average electro-magnetic torque produced over a rotor movement. From this comparison, it is seen that the SRM has uni-polar mmf; phase-flux variations and energy conversion loop are limited to the first quadrant only whereas the DSPM has bipolar mmf and uni-polar phase-flux, with the energy conversion loop limited to two quadrants. However, the FRM has bipolar mmf and phase-flux variation with respect to rotor movement, the energy conversion loop covers the four quadrants; meaning that control is possible for all four quadrants. The typical 2D and 3D cross-section of FRM of 6/8 poles are illustrated in Figure 17. The flux and flux density distribution in the machine at aligned and un-aligned positions of the rotor are illustrated in Figure 18. The variation of mmf and phase-flux with respect to rotor displacement are shown in Figure 19. The principle of FRM is variable flux linkages, inducing an emf that interacts with alternating armature current as seen in Figure 18. The ideal variation of mmf and phase flux with respect to the rotor displacement is seen in Figure 19. The field excitation provided by the PMs, armature winding flux linkages are modulated by the variation of the magnetic circuit reluctance, as rotor displaces, in such a way that bipolar induced emf are produced without rotating the PMs [15].

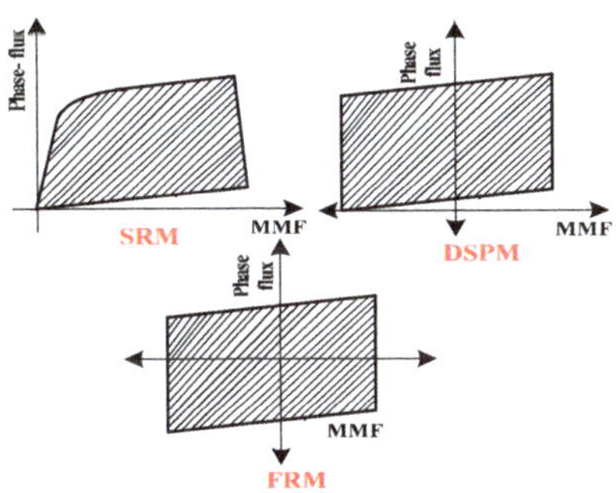

Figure 16. Comparisons with three double salient type machines [14].

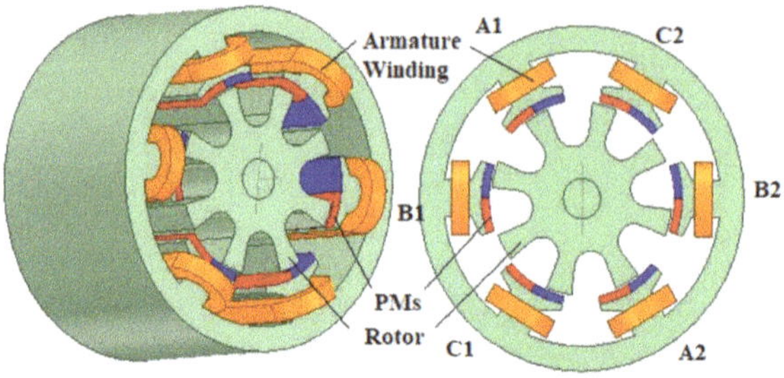

Figure 17. A typical 3D and 2D structures of designed FRM.

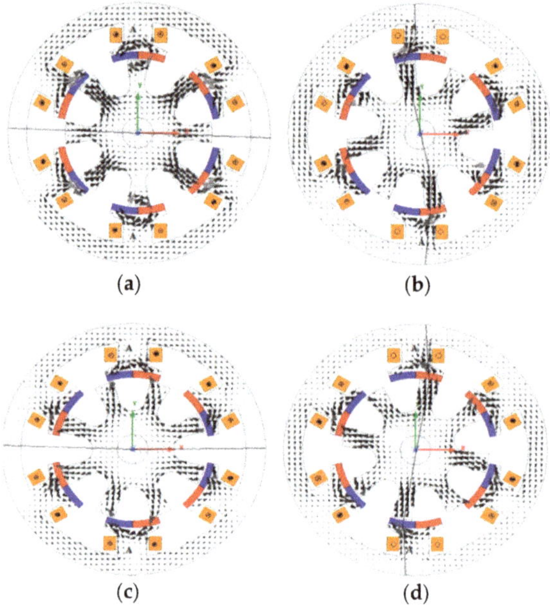

Figure 18. Flux-distribution of six/eight-pole FRM: (**a**) rotor position at 0°; (**b**) rotor position at 11.25° ACWD; (**c**) rotor position 22.5° ACWD; (**d**) rotor position at 11.25° CWD.

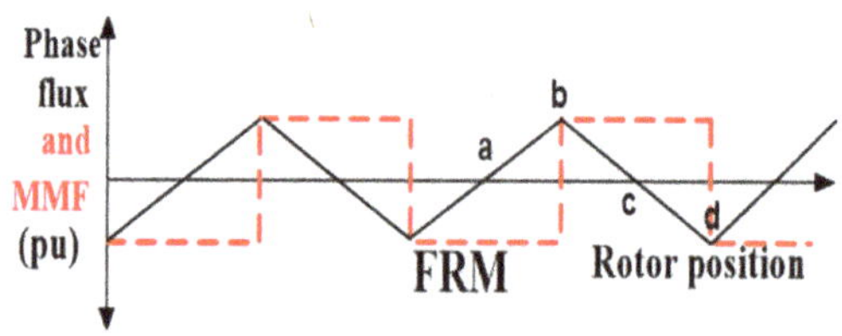

Figure 19. Variation of mmf and phase-flux variations with respect to rotor displacement [15].

In the generator case, when the rotor is driven by the prime mover as in Figure 18a, the equilibrium position of the rotor poles show they are unaligned with stator poles. At that position, there are no flux linkages with the coils. Only the flux setup by the magnets circulates completely within each stator teeth (point "a" in Figure 19). In Figure 18b, the rotor is driven to 11.25° in an anti-clockwise direction (ACWD) then the rotor poles overlap with the stator pole magnets and flux creates the path from upper PMs, lower PMs, and stator back core iron. Phase flux is extreme in this position at (point "b" in Figure 19). In Figure 18c, again at equilibrium position which is displaced from the first equilibrium position by 22.5° no flux links with the coils (point "c" in Figure 19). Further movement of the rotor leads to the position in Figure 18d of 11.25° clockwise direction (CWD), where phase flux is again extreme in the opposite direction to that of the first alignment position (point "d" in Figure 19). Linear bipolar phase flux variation induces a total induced emf, e_O. According to the Faraday's law of electromagnetic induction (1), a change in flux linkage produced by ψ_m (field source), at a given electrical speed ω_e in rad/s, the emf induced e_O is given as

$$e_O = -\omega_e \frac{d\psi_m}{d\theta} \tag{1}$$

Table 2 shows the evaluated dimensions of a FRM through the sizing equation technique [23]. Based on these dimensions, FRM has been analyzed in 2D FEA and by using the magneto-static solver analyzed the flux density distribution and cogging torque. The meshed plot and no-load flux-density distribution of the modeled machine are shown in Figure 20. The magneto-static cogging torque is shown in Figure 21. It is important to note that FRM exhibits fault tolerance ability because the mutual inductance magnitude value is less than one-fiftieth of the self-inductance of the phases which indicates that natural isolation between the phases. The variation of the inductances with respect to the rotor displacement is small. Here the mutual inductance 4 mH and self-inductance is 0.61 H. Therefore, the reluctance torque is insignificant.

Table 2. Dimensional details of non-rare-earth (NRE) FRM.

S. No	Parameters	Value	Symbol
Stator	magnet thickness	3 mm	h_{pm}
	stator-pole span angle	42.6°	β_s
	stator-pole height	15 mm	h_{ps}
	outer diameter of stator	129	D_s
	stator arc span	27.8 mm	τ_{ps}
	stack length	400 mm	l_{stk}
	no. of turns/coil	176	N_{ph}
	air gap	0.5 mm	g
	magnet remanence	0.4T (Sr-Fe)	0.4T (Sr-Fe)
Rotor	outer diameter of rotor	72 mm	D_r
	rotor arc span	12.3 mm	τ_{pr}
	rotor pole span angle	21°	β_r
	rotor pole height	17 mm	h_{pr}

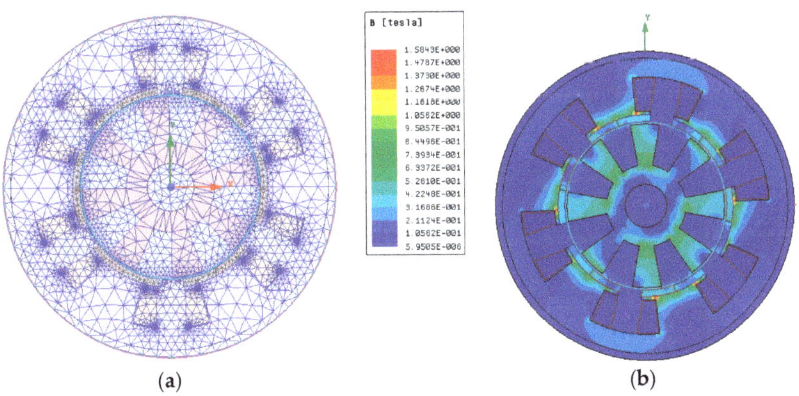

Figure 20. FEA evaluation of no-load behavior of FRM: (**a**) mesh plot; (**b**) flux density map.

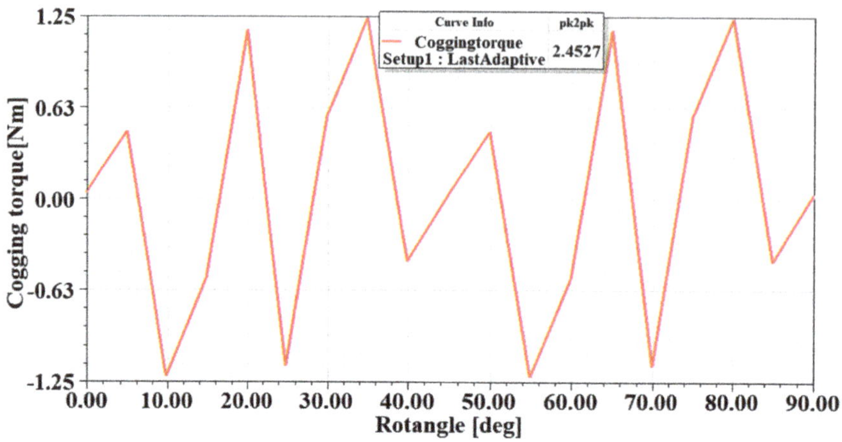

Figure 21. Cogging torque cycle under magneto-static no-load (MSNL).

2.3. Design Topologies

A general rule regarding the number of rotor poles N_r and the number of stator poles N_s and choosing "p" for the number of phases of FRM is [17]:

$$\frac{N_s}{N_r} = \frac{p}{p+1} \tag{2}$$

Hence N_s/N_r= 3/4, 6/8, 12/16, etc., are prevalent for 3-phase FRM machines (2).

If a number of pairs of PMs are given, then [23]:

$$N_s\left(\frac{n_p+1}{3}\right) = N_r \tag{3}$$

where n_p = number of PMs pair poles. Various possible configurations of the FRM are shown in Table 3. For 6/8 FRM configuration accompanied one pair of PMs in each stator pole (3). The no-load flux density plot of 6/8 pole FRM is illustrated in Figure 20. In addition, the number of possible PMs on the stator surface and the number of effective poles corresponding to the flux patterns are given in Table 3. Table 3 shows that as the number of rotor poles increases, speed decreases. Increasing the rotor pole number means increasing the energy cycles per revolution. Thus, the reluctance torque is negligibly small, even though the CT is still slightly high for LS applications. With the proper design and control, the cogging torque component can be made relatively small. The number of poles corresponding to the effective flux patterns, speed, and gear ratio for different FRM topologies for LS applications are given in Table 3. A summary of the various FRM topologies highlighted so far is evaluated from ten points of view with reference to wind turbines in Table 4. The performance marks are given in a percentage scale. In Table 4, these percentage values are represented in form of their fractional values.

Table 3. Various possible configurations of PM FRM.

Machine Type	Number of Effective Flux Patterns	Number of Magnets	Speed for 50 Hz (r/min)	Cogging Torque Cycle
6/8	2	12	375	15°
6/14	2	24	214	8.57°
12/16	4	24	188	7.5°
12/10	4	24	300	6.0°
12/28	4	48	108	4.2°
12/40	4	60	75	3.0°
48/46	4	96	65.3	0.33°

Table 4. Comparisons of various topology rare-earth (RE)-PM FRMs for wind energy applications.

Properties	Figure 4 [15]	Figure 5 [16,17]	Figure 6a [23]	Figure 6b [23]	Figure 7 [24]	Figure 8 [24]	Figure 9 [25,26]	Figure 10a [27,28]	Figure 10b [27,28]	Figure 11 [29]	Figure 12 [31]	Figure 13 [32]	Figure 14 [32]	Figure 15 [32]
Constructional simplicity	1	1	1	0.8	0.8	0.6	0.4	0.4	0.4	0.8	0.1	0.8	0.8	0.6
Power Density	0.6	0.8	0.8	0.8	0.8	1	1	0.8	1	0.8	0.6	0.6	0.8	0.8
Usage of Rotor Volume	0.8	0.8	0.8	0.6	0.8	0.8	0.6	0.8	1	0.8	0.8	0.4	0.8	1
Low material& manufacturing cost	1	0.8	0.6	0.4	0.8	0.6	0.6	0.4	0.4	0.8	1	0.6	0.6	0.6
Low CT	0.8	0.8	0.6	0.6	0.8	0.8	0.8	0.6	0.6	0.8	0.8	0.4	0.4	0.6
High Temperature operation	0.8	0.8	0.8	0.6	0.8	0.8	0.8	0.8	1	0.8	0.8	0.6	0.6	0.8
Fault tolerance	1	0.8	0.8	0.6	0.8	0.8	0.8	1	0.8	1	0.6	0.6	0.8	0.8
Easy replacement of faulted coils	0.8	0.8	0.6	0.6	0.8	0.6	0.6	0.8	0.8	0.8	0.6	0.6	0.8	0.6
Mechanical rigidity	0.8	0.8	0.6	0.6	0.8	0.6	0.8	0.6	0.6	0.8	0.6	0.6	0.6	0.8
Flux Leakage	0.8	0.8	0.6	1	0.8	0.8	0.8	0.8	0.8	0.8	0.6	0.6	0.6	1

3. Analytical Calculation of Cogging Torque and Torque Ripple

The 2-dimensional governing expression of PM FRM can be expressed in magnetic vector potential A as [34]:

$$\frac{\partial}{\partial x}\left(\frac{1}{\mu}\frac{\partial A_z}{\partial x}\right)+\frac{\partial}{\partial y}\left(\frac{1}{\mu}\frac{\partial A_z}{\partial y}\right)=-J_o-\left(\frac{\partial M_y}{\partial x}-\frac{\partial M_x}{\partial y}\right) \tag{4}$$

where A_z = vector potential in z, J_o = current density, M = magnetization of PM.

The flux linkage λ are calculated per phase from the average vector potential 'A' (4) over each winding cross-section as

$$\lambda=\left(\iint_{s_1} A_1\frac{ds_1}{s_1}-\iint_{s_1} A_2\frac{ds_2}{s_2}\right)l_{stk} \tag{5}$$

where l_{stk} is stack length; s_1 and s_2 are the total area of N_{ph}-turns per phase with the winding carrying the negative and positive currents, respectively. The flux linkages of each phase of 3-phase FRM machines can be evaluated (5).

FRM mainly suffers from unfavorably large CT due to its double-salient construction and PMs, resulting in undesirable vibration and noise especially for LS and MS wind energy applications. At high speed, this effect is negligible [18]. The CT does not add to the average electro-magnetic torque, but only affects the pulsations in the torque. It is also known as self-aligning torque or no-current torque. It occurs mainly in PM machines. It is the torque due to the interaction between the PM and the stator slots. When there is no current in stator windings, as the saliency nature of rotor and stator with uneven permeance in the air gap produces the cogging torque [18]. The no-load open-circuit flux density distribution is obtained from Equation (6), based on the mmf-permeance model in the airgap [19,20].

$$B(\alpha,\theta)=F_{pm}(\alpha)\wedge(\alpha,\theta) \tag{6}$$

where $\wedge(\alpha,\theta)$ and $F_{pm}(\alpha)$ are permeance in the air gap and the mmf excited by PMs, respectively. α is the rotor position along the air gap circumference, θ is rotor position, w is the total energy stored in the air gap and υ is the volume of the air-gap.

The analytical expression of CT of FRM machines can be obtained [20] from the instantaneous torque for every cycle of rotor displacement by means of the rate of change of co-energy with respect to the rotor movement as shown in Equation (7).

The 2D FEA is used to predict the overall CT as

$$\begin{aligned} T_{cog} &= -\tfrac{\partial w}{\partial\theta}=-\tfrac{\partial}{\partial\theta}\left(\tfrac{1}{2\mu_o}\int B^2(\alpha,\theta)\right)d\upsilon \\ &= \tfrac{\partial}{\partial\theta}\left(\tfrac{1}{2\mu_o}\int F_{pm}^2(\theta)\wedge^2(\alpha,\theta)d\upsilon\right) \end{aligned} \tag{7}$$

The cogging torque is calculated when the machine is at a standstill position and when there is no current in the coils of the stator (i = 0). It is supposed that each section plane along axial direction has the same magnetic field distribution (6). The cogging torque of FRM is obtained by having the calculated value multiplied by the total machine length or stack-length [18]. The overall cogging torque mainly depends on airgap permeance and PMs mmf functions (7).

The analytical expression of total cogging torque for conventional stator active PM brushless FRM can be expressed as [20]:

$$T_{cog}=\frac{\left(R_2^2-R_1^2\right)L_aN_r\pi}{4\mu_o}\sum_{m=1}^{\infty} mF_{pm(mN_r/N_s)}\wedge_m sin(mN_r\theta) \tag{8}$$

where L_a= effective stack length, R_1 and R_2 are the outer rotor radius and inner stator radius, $\wedge_m$ is Fourier coefficient, and m is satisfied as the following expression is given as

$$m = \frac{kN_s}{GCD(N_s, N_r)} \text{ k } = 1, 2, 3 \ldots.. \tag{9}$$

where $GCD(N_s, N_r)$ is the highest common factor of N_s and N_r. The Equations (8) and (9) can be appropriate for FRMs of any slot-pole configurations [20].

The torque ripple factor k_{rp} is evaluated as

$$k_{rp} = \left(\frac{T_{max} - T_{min}}{T_e} \times 100 \right) \% \tag{10}$$

3.1. Cogging Torque Reduction Techniques for PM FRM

Many researchers studied and developed different design techniques to minimize the CT of PM machine structures [19]. The most accepted techniques to minimize CT is to optimally vary the dimensions of the machine design such as rotor skewing, rotor teeth pairing, chamfering the rotor and stator PM thickness, and rotor geometrical modifications. Generally, stator side modifications consist of tooth pairing, magnet skewing, slot shifting, and dummy slots [35,36]. The stator side modifications augment the machine size and cost. Thus, the stator modifications are not practical in the case of PM machines. Then, the rotor side modifications are skewing rotor, chamfering, and radial pole pairing, and axial pole pairing [18–20]. The required modifications in the rotor than in the stator are generally used for minimizing cogging torque. Therefore, in this paper, the analysis part is divided into three subsections, where the first and second are used to report analysis of the existing CT minimization techniques like skewing and rotor pole pairing (CPOP). In the third subsection, a novel model constituting, the combining of these two existing techniques (SKCpp) is additionally investigated. Compared are different methods for electromagnetic performances of a generator with various rotor side design modifications.

3.1.1. Skew Rotor Design

Skewing is the most renowned and widely used technique to minimize the CT effects in PM machines. Skewing can be done in either stator or rotor. In the FRM, the stator has magnets while the rotor is composed entirely of laminations. So, rotor skewing is more feasible. Furthermore, when the rotor is skewed with the optimal skewing angle, the CT is effectively minimized. The skewed rotor yields even permeance in between the stator and rotor irrespective of the rotor displacement [17]. In this study, slicing or stepped skew technique has been incorporated into the rotor structure and the rotor is sliced into five segments along the length. For various skew angles (5°, 10°, 15°, and 20°), the CT effect is analyzed by 2D FEA as shown in Figure 22. The CT variation with rotor position has a cycle of $2\pi/N_s$. The CT cycle is 15°, which agrees with FEA simulation as in Figure 22. Based on the simulation 15° skew angle effectively minimized the CT compared with the other angles. FEA shows that the CT is decreased by 92.2%. Nevertheless, this method leads to some reduction of flux linkages, induced emf, and the corresponding reduction of power density.

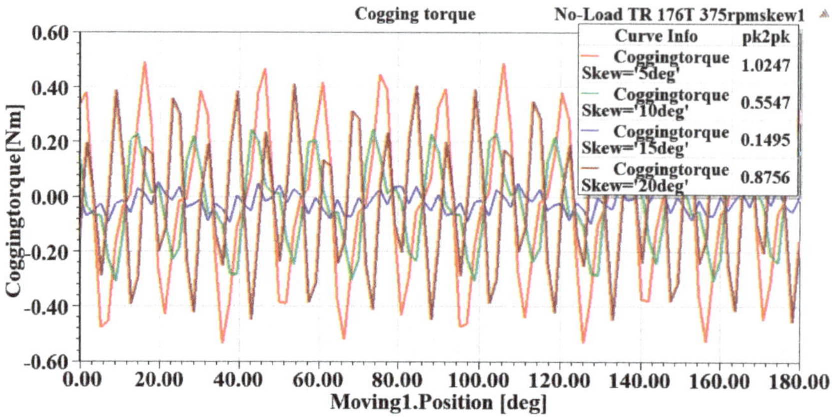

Figure 22. The effect of cogging torque for various skew angles.

3.1.2. Circumferential Rotor Teeth Pairing (β_r) Method or Rotor Pole Pairing (CPOP)

The rotor teeth pairing method is also considered to minimize the CT of FRM. The CT waveform varies with the rotor tooth width β_r. This method of circumferential or rotor teeth pairing method employs two different pole widths in the rotor design that can be applied. The variable magnetic reluctance of the rotor and the air gap minimizes the amplitude of the CT. Based on the Fourier series expansion [18,20], by adjusting the rotor pole width with respect to the PM width with these combinations the optimal rotor tooth width has to be selected as 19° and 21°. The rotor poles are designed into two different pole arc widths, these two arcs are oppositely employed as illustrated in Figure 23. The overall cogging torque got reduced as verified by FEA. This technique reduced the CT by 14% of the original value in transient no-load. The overall CT can also be minimized.

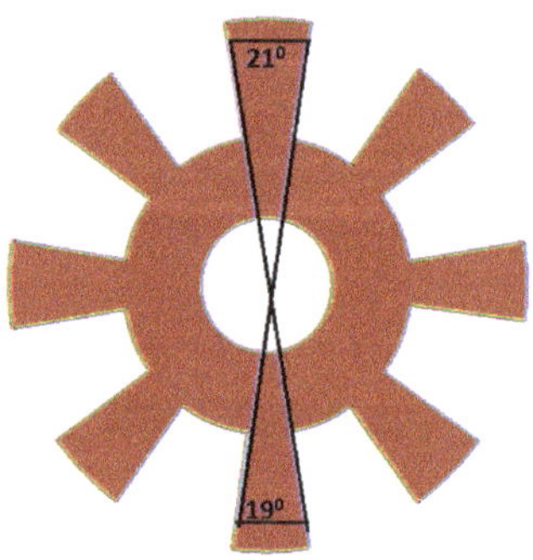

Figure 23. Cross-section of teeth paired rotor.

3.1.3. Combined Auxiliary Model of FRM

The skewing and rotor pole pairing methods are explained above; this paper attempts to combine these two methods to obtain the performance of FRM better than any one of the two. This combined method is the possible combinations of more than two existing CT reduction methods. There are various combined methods are studied for PM machines. For example, the combination of rotor magnet skewing with teeth notching by B. Zhang [21] and the combined pole and slot number by F. Yusivar [37]. The 3D structural view of SKCpp is depicted in Figure 24. This method of combining the existing two methods is a novel method and is being introduced for the first time. By this method, the CT is reduced by 94.8% of the original value in Transient (TR) no-load by 2D FEA. Compared to the conventional skewing, rotor teeth-pairing, the SKCpp also effectively minimized the cogging torque while maintaining all other electromagnetic performances as in the basic model.

Figure 24. 3D Structural view of SKCpp.

3.2. Basic Performance Evaluation

Some evaluations are carried out to examine the generator performances for each rotor design in terms of flux linkages, induced voltages, and cogging torque.

3.2.1. Flux Linkage

Under the open-circuit condition, the flux from PMs is investigated with Equation (5). A 180° rotation of the rotor, with step time as 0.1 ms and simulated time of 80 ms results in four cycles of 50 Hz frequency and 800 simulations. Figure 25 clearly shows that CPOP has the highest flux-linkages amplitude that emerges from the PM compared to other generators, while skew has the lowest flux linkage compared to the other generators. For basic, skew, CPOP, and combined auxiliary models (SKCpp), the measured maximum flux linkages are 0.41 Wb, 0.327 Wb, 0.433 Wb, and 0.348 Wb, respectively.

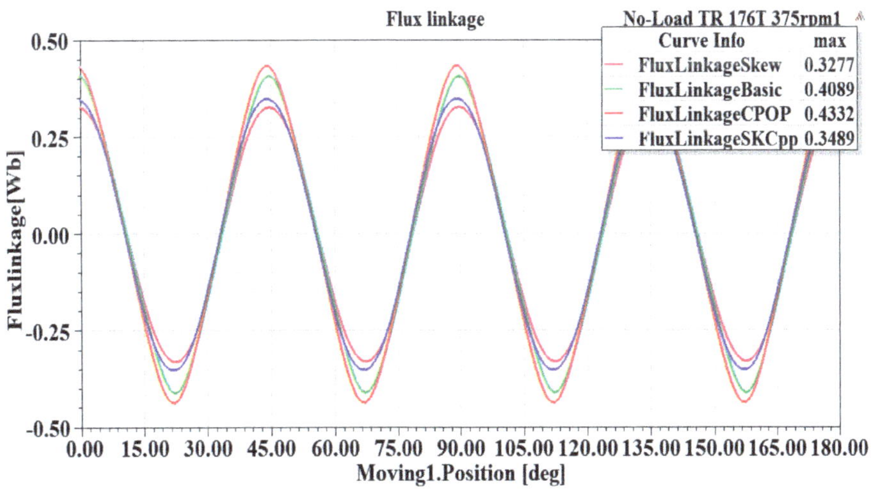

Figure 25. Flux linkage waveforms comparison.

3.2.2. Induced Voltage

Further, under open circuit condition (TR-no-load), the rotor is rotating at a prescribed speed of 375 r/min. The induced emfs of the four-rotor designs are analyzed and compared in Figure 26.

As expected, the highest magnitude is measured with CPOP as 118.4 V followed by basic, SKCpp and skewing with 112 V, 109.5 V, and 101.7 V, respectively.

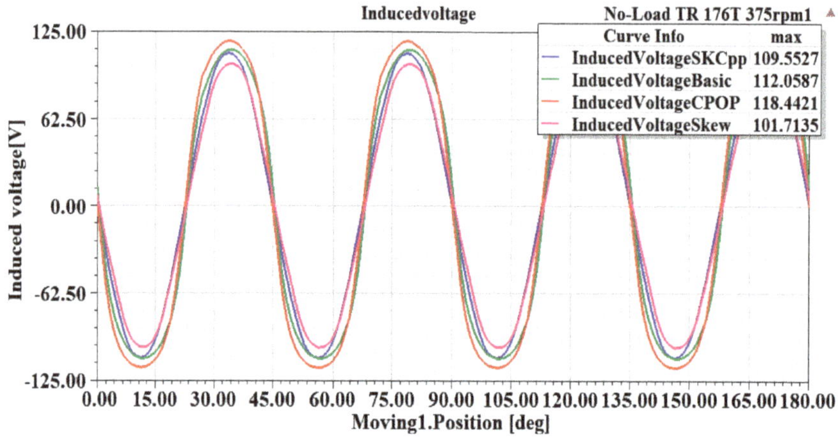

Figure 26. Induced voltage waveforms comparison at transient no-load.

3.2.3. Cogging Torque (CT) Reduction

The CT as applied to PM machines is the torque effort due to magnets only viz., no load current present. The CT leads to poor position control, vibration, noise, and reduction in generator performance. By setting rotor speed at 375 r/min, and one complete electrical cycle of 180°, CT for various rotor configurations is plotted as shown in Figure 27. This shows the conventional (basic) rotor model has the highest peak-to-peak CT value of 1.82 Nm followed by the values for skewed, rotor teeth pairing (CPOP) and SKCpp models as 0.14 Nm, 1.56 Nm, and 0.09 Nm, respectively. As compared to the basic model, skew and SKCpp gives that highest CT reduction. In Table 5, the reduction in CT at TR-no-load is compared quantitatively, which indicates that the generator acoustic noise is lowest with proposed SKCpp.

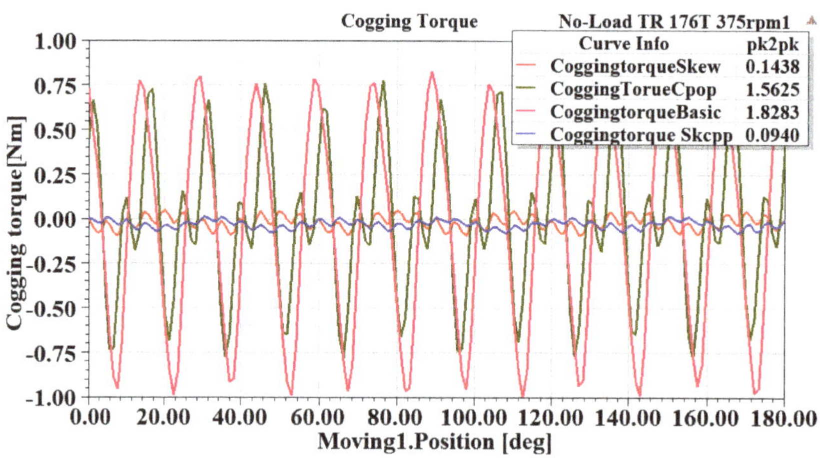

Figure 27. Comparisons of cogging torque effect for various methods at transient no-load.

Table 5. Comparison table of reduction of CT for NRE PM FRM at TR-no-load.

Rotor Model	Cogging Torque (Nm)	Cogging Torque Reduction (%)
Conventional (6/8)	1.283	0
Skew	0.1438	92.2
CPOP	1.5625	14.5
SKCpp	0.0940	94.8

3.2.4. Torque Ripple Reduction Analysis (TR-Load)

The characteristics of the torque ripple lead to pulsating torque which results in noise and vibration, impacting negatively on the machine performance under load. The torque ripple factor is evaluated by Equation (10) and the torque ripple for different rotor modification techniques at the rated load is evaluated as shown in Figure 28. It shows that the basic model has the highest torque ripple factor value of 18.6 % followed by the values of skewing, CPOP, and SKCpp at 6.73%, 17.4%, and 5.5%, respectively. In comparison to the basic model, the CPOP, and skew, the SKCpp model shows the highest torque-ripple reduction. In Table 6, the reduction in torque ripple at TR-load is compared quantitatively with the conventional design as the reference.

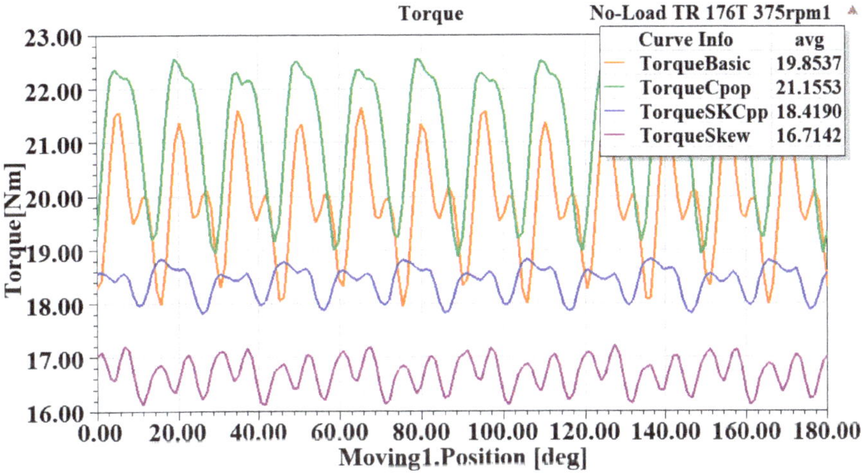

Figure 28. Comparisons of a torque ripple effect for various methods at transient load.

Table 6. Comparison table of reduction of torque ripple for NRE-PM FRM at TR-rated-load.

Rotor Model	Average Torque, T_e (Nm)	Torque Ripple Reduction (%)
Conventional (6/8)	19.8537	0
Skew	16.7142	64
CPOP	21.1553	8
SKCpp	18.419	71

3.3. Power Generating Performance Comparison

The generating operating point performances including output voltage against load current, voltage regulation, power curve in terms of varying load current and generator speed, as well as loss and efficiency curves of the four generators working are analyzed by FEA simulation. The four generators are designed to work under symmetrical resistive loads. Hence, based on the rated voltage of each phase and rated power, the calculated rated load resistance, R_n, is 25.3 Ω.

Figure 29 illustrates the variation curves of the output voltage versus phase current—so-called overload capability curves—while the voltage regulation is presented in Figure 30. One can see from Figure 29 that the CPOP is presented with as the machine with the highest overload capability, while basic, SKCpp and skew follow in that order. In Figure 30, as load current increases, the skew is presented with the highest voltage regulation, followed by SKCpp, basic, and CPOP. Lower percentage values in Figure 30 is an indication that there is a less voltage variation when the load changes in the generators.

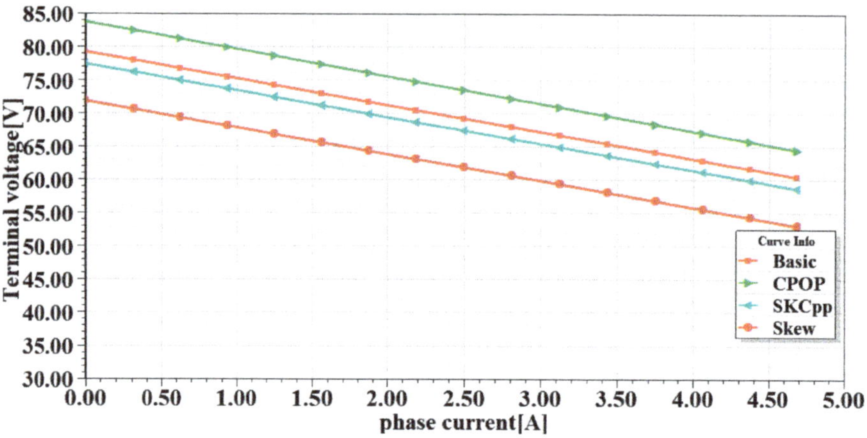

Figure 29. Output voltage against phase current @ 375 r/min.

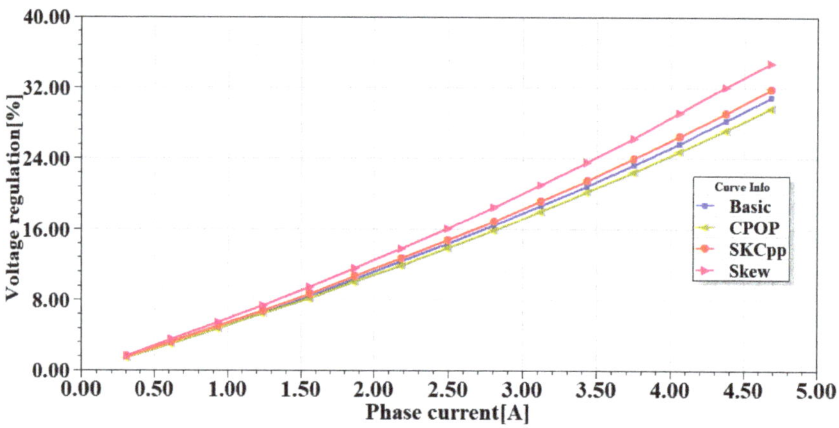

Figure 30. Voltage-regulation against phase current (rms) for each rotor designs @ 375 r/min.

The generator load profiles are exhibited by varying the phase current against the output power for each rotor design as shown in Figure 31. The overload limits are (4.6875 A, 850.77 W), (4.6875 A, 747.42 W), (4.6875 A, 907.611 W) and (4.6875 A, 825.78 W) for basic, skew, CPOP, and SKCpp, respectively, which is up to 1.5 times of the rated load. Meanwhile, the efficiencies are compared in Figure 32, for all generators.

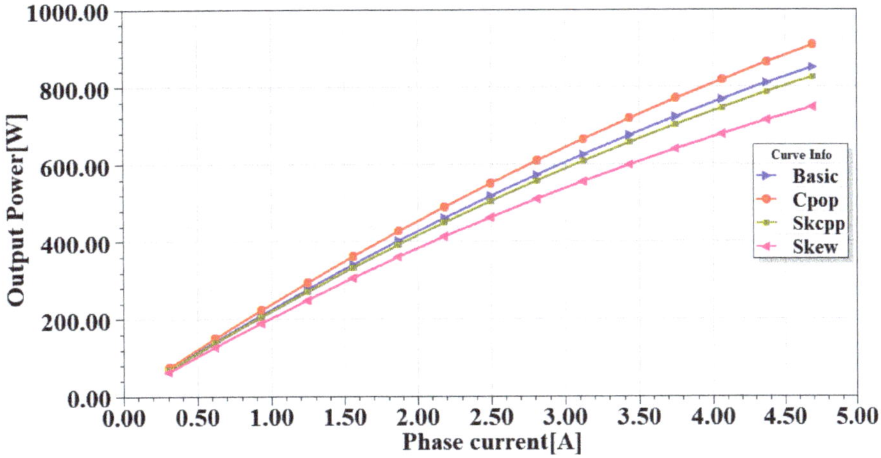

Figure 31. Output-power against phase current (rms) for each rotor designs @ 375 r/min.

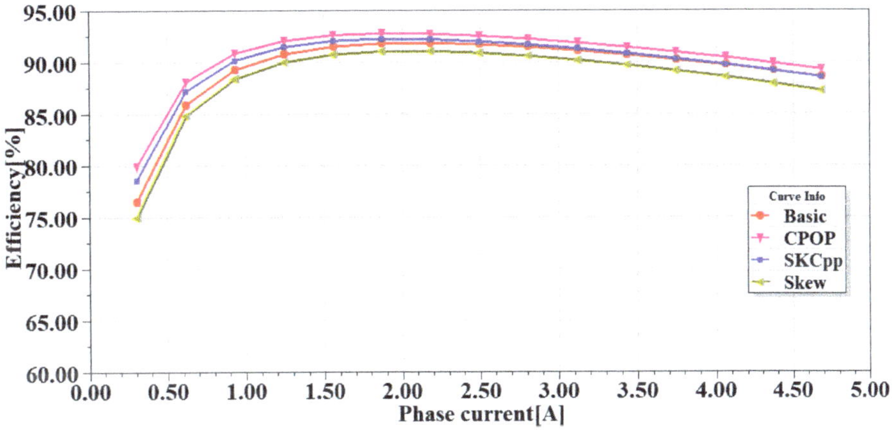

Figure 32. Efficiency against phase current (rms) for each rotor designs @ 375 r/min.

In terms of speed capability, the open circuit EMF increases with speed, which translates to increasing power density. The speed capability of the generators are analyzed when connected with a symmetrical rated load while varying the mechanical speed as shown in Figure 33. The output power of the skewed rotor generator becomes easily saturated and decreases with the increase in the speed of the rotor. The core loss and efficiency curves against rotor speed are also plotted as shown in Figures 34 and 35, respectively. As expected, an increase in speed, increases the core-loss magnitudes. As the rotor speed increases, the core losses become distinct and it is observed that SKCpp presents the lowest value compared with the other generators.

Overall, from the 2D transient FEA simulations, a summary of the performance for all four generator designs is provided as shown in Table 7. It is interesting to note that the proposed SKCpp design yields comparatively improved performance characteristics in terms of cogging torque and torque ripple reduction, as well as efficiency, while CPOP presents better performance with respect to stability for overload capability and voltage.

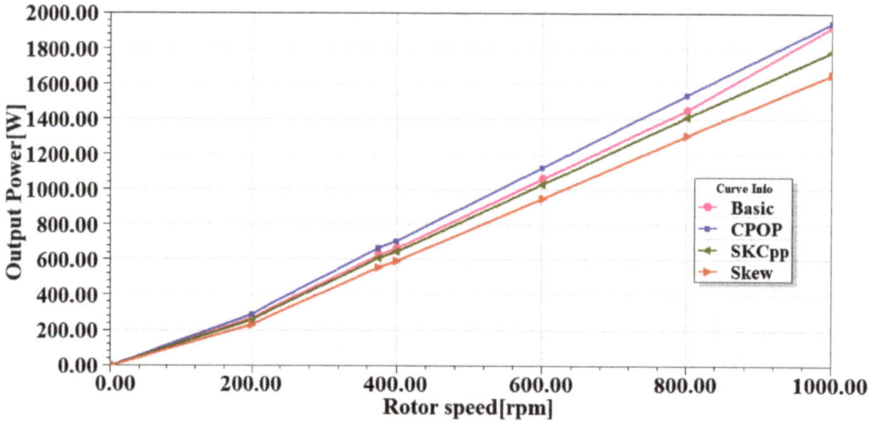

Figure 33. Output-power against mechanical speed (rpm) for each rotor design @ 375 r/min.

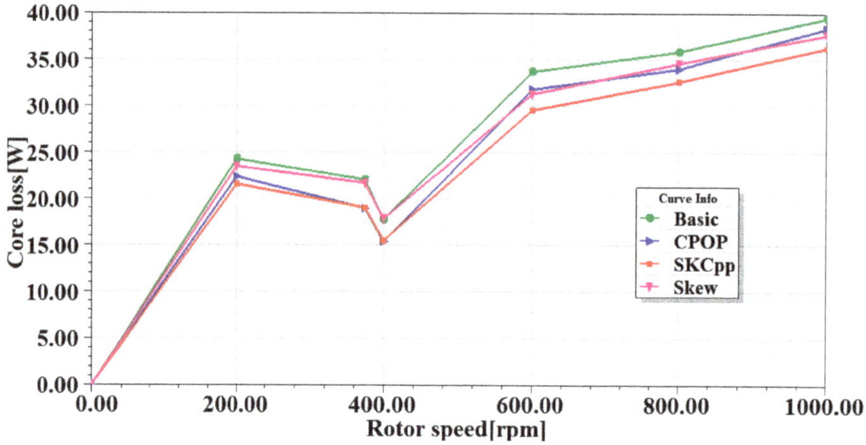

Figure 34. Core-loss against rotor mechanical speed (rpm) for each rotor design @ 375 r/min.

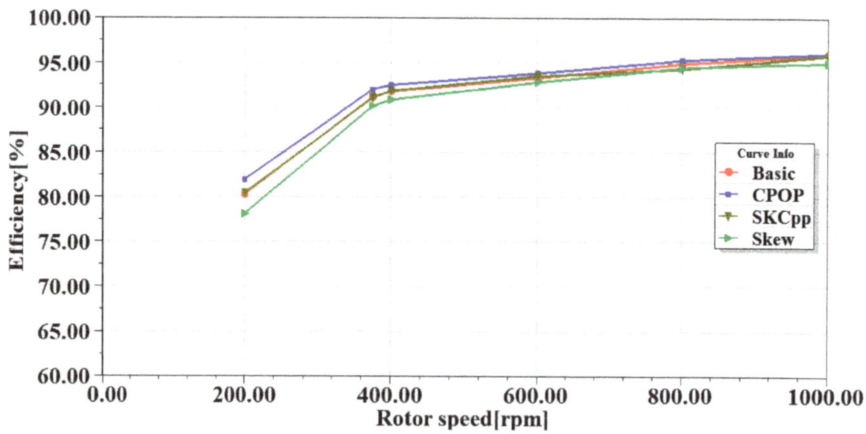

Figure 35. Efficiency against rotor mechanical speed (rpm) for each rotor design @ 375 r/min.

Table 7. Comparison table of all the rotor designs of 6/8 NRE-PM FRM Generators.

Rotor Model	Cogging Torque	Torque Ripple	PowerLimit	Voltage Regulation	Efficiency
Basic (6/8)	Poor	Poor	Moderate	Moderate	Good
Skew	Moderate	Moderate	Poor	Poor	Moderate
CPOP	Poor	Poor	Good	Good	Good
SKCpp	Good	Good	Moderate	Moderate	Good

4. Conclusions

In this study, the possibility of using NRE (ferrite) PM excitations to design the flux reversal machine for direct drive wind turbine generator has been demonstrated for the first time. Various existing RE PM-FRM structures have been reviewed. In this frame of overview, fourteen RE PM FRM designs were highlighted and examined in detail for their suitability for wind energy generation. The merits and demerits of these FRM topologies were quantitatively deliberated in terms of power density, cogging torque, and torque-ripple.

Thereafter, a 6/8 pole ferrite PM FRM is redesigned with a conventional rotor model, as well as with skewing and rotor pole pairing, while a new combination of skewing with circumferential rotor pole pairing (SKCpp) is also designed and proposed for minimizing cogging torque and torque ripple. The machines are numerically evaluated for wind power generation in 2D FEA. In addition, the generator performance, in terms of flux linkages, induced voltages, over-load and over-speed capabilities, as well as efficiency, are investigated and discussed. The results obtained through 2D FEA simulations show that the proposed SKCpp technique as the best design approach for reducing the cogging torque and torque ripple effect in ferrite PM FRM, with a percentage reduction of 94.8% at no-load and 71% under load, respectively. However, the average torque is reduced by 13%. Overall, the potential of the proposed ferrite PM FRM for wind power generation is clearly demonstrated for the reduction of cogging torque and torque ripple and also for efficiency, overload, and speed regulation.

Author Contributions: Conceptualization, B.M. and U.B.A.; methodology, B.M. and U.B.A.; software, B.M.; validation, B.M. and U.B.A.; formal analysis, B.M. and U.B.A.; investigation, B.M.; resources, M.K.K.; data curation, B.M.; writing—original draft preparation, U.B.A. and B.M.; writing—review and editing, B.M., U.B.A. and M.K.K.; visualization, B.M.; supervision, U.B.A. and M.K.K.; project administration, B.M.; funding acquisition, U.B.A. All authors have read and agreed to the published version of the manuscript.

Funding: This research received no external funding but the APC is funded by Tshwane University of Technology, South Africa.

Conflicts of Interest: The authors declare no conflict of interest.

Nomenclature

Symbol	Stands for
e_O	Induced EMF (V)
ω_e	Electrical speed (rad/s)
h_{pm}	PM thickness (mm)
τ_{ps}	Stator arc span (mm)
h_{ps}	Height of stator pole (mm)
g	Length of the airgap (mm)
l_{stk}	Length of stack laminations (mm)
D_s	Outer diameter of stator (mm)
h_{pr}	Height of rotor pole (mm)
β_r	Span angle of rotor pole (° mech.)
β_s	Span angle of stator pole (° mech.)
D_r	Outer diameter of rotor (mm)
L_a	Effective stator stack length
N_{ph}	Number of turns per coil
m	Number of phases

N_r	Rotor pole number
ψ_m	The field source Flux linkage (Wb)
N_s	Stator pole number
n_p	Number of PMs Pairs
T_{cog}	Cogging torque (Nm)
A	Magnetic vector potential (T-m)
J_o	Current density per phase (A/mm^2)
M	Magnetization of PMs (A/mm)
τ_{pr}	Rotor arc span (mm)
λ	Flux linkages per phase (Wb-t)
θ	Position of the rotor (° mech.)
w	Magnetic co-energy (J)
F_{pm}	MMF due to Permanent magnets (A-t)
B	No-load flux density (T)
$\wedge$	Permeance (Wb/At)
υ	Air-gap volume (mm^3)
μ_o	Permeability of free space (H/m)
s	Area of N_{ph} turns
R_1	Outer rotor radius (mm)
R_2	Inner stator radius (mm)
k_{rp}	Ripple-torque factor (%)
α	Position along the circumference in the airgap
T_{max}	Maximum Torque (Nm)
T_{min}	Minimum Torque (Nm)
T_e	Average Torque (Nm)

Abbreviations

PM	Permanent magnet
RE	Rare-earth
NRE	Non-rare earth
HS	High speed
LS	Low speed
FRM	Flux reversal machine
FSM	Flux switching machine
SMPM	Stator-mounted permanent magnet machine
CT	Cogging torque
FMDT	Flux-mmf diagram
FEA	Finite element analysis
PMSM	Permanent magnet synchronous machine
FSCW	Fractional slot concentrated winding
FPW	Full pitched winding
CW	Concentrated winding
CPM	Consequent-pole permanent magnet
TFPMLM	Transverse-flux permanent magnet linear machine
SRM	Switched reluctance machine
DSPM	Double salient permanent magnet machine
TR	Transient
CWD	Clockwise direction
ACWD	Anti-clockwise direction
CPOP	Circumferential rotor pole pairing
SKCpp	Skewing with circumferential pole pairing

References

1. REN 21. Renewables 2020 Global Status Report (Paris: REN21 Secretariat). Available online: http://www.ren21.net/gsr (accessed on 10 October 2020).
2. Akuru, U.B.; Kamper, M.J. Evaluation of flux switching PM machines for medium-speed wind generator drives. In Proceedings of the IEEE Energy Conversion Congress and Exposition (ECCE), Montreal, QC, Canada, 2015; pp. 1925–1931.
3. Thomas, J. Getting rare-earth magnets out of EV traction machines: A review of the many approaches being pursued to minimize or eliminate rare-earth magnets from future EV drive trains. *IEEE Electrif. Mag.* **2017**, *5*, 6–18.
4. Arshad, S.; Selvi, V. 2D electromagnetic design of flux reversal generator for low speed wind applications using finite element analysis. *Int. J. Eng. Technol. Sci. Res.* **2017**, *4*, 149–157.
5. More, D.S.; Kalluru, H.; Fernandes, B.G. Comparative analysis of flux reversal machine with fractional slot concentrated winding PMSM. In Proceedings of the 34th Annual Conference of IEEE Industrial Electronics, Orlando, FL, USA, 10–13 November 2008; pp. 1131–1136.
6. More, D.S.; Fernandes, B.G. Analysis of flux-reversal machine based on fictitious electrical gear. *IEEE Trans. Energy Convers.* **2010**, *25*, 940–947. [CrossRef]
7. Pellegrino, G.M.; Gerada, C. Modeling of flux reversal machines for direct drive applications. In Proceedings of the 14th European Conference on Power Electronics and Applications, Birmingham, UK, 30 August–1 September 2011; pp. 10–18.
8. Akuru, U.B.; Kamper, M.J. Investigation of low-cost PM flux switching machine for medium speed geared wind energy application. In Proceedings of the 25th South African Universities Power Engineering Conference (SAUPEC), Stellenbosch, South Africa, 2017; pp. 613–618.
9. Akuru, U.B. Design Optimization and Performance Evaluation of Flux Switching Machines for Geared Medium-Speed Wind Generator Drives. In Proceedings of the IEEE PES/IAS Power Africa Conference, Cape Town, South Africa, 2018; pp. 925–930.
10. Yuting, G.; Ronghai, Q.; Dawei, L.; Li, J. Torque performance analysis of three-phase flux reversal machines for electrical vehicle propulsion. In Proceedings of the 2016 IEEE Transportation Electrification Conference and Expo Asia-Pacific (ITEC Asia-Pacific), Busan, Korea, 1–4 June 2016; pp. 297–301.
11. Liao, Y.; Liang, F.; Lipo, T.A. A novel permanent magnet motor with doubly salient structure. *IEEE Trans. Ind. Appl.* **1995**, *31*, 1069–1078. [CrossRef]
12. Jianzhong, Z.; Ming, C.; Zhe, C.; Wei, H. Comparison on stator-mounted permanent-magnet machines based on a general power equation. *IEEE Trans. Energy Convers.* **2009**, *24*, 826–834. [CrossRef]
13. Rauch, S.E.; Johnson, L.J. Design principles of flux-switch alternators. *IEEE Trans. Am. Inst. Electr. Eng.* **1955**, *74*, 1261–1268.
14. Deodhar, R.P. The Flux-MMF Diagram Technique and Its Applications in Analysis and Comparative Evaluation of Electrical Machines. Ph.D. Thesis, University of Glasgow, Glasgow, UK, October 1996.
15. Deodhar, R.P.; Andersson, S.; Boldea, I.; Miller, T.J.E. The flux reversal machine: A new brushless double-salient permanent magnet machine. *IEEE Trans. Ind. Appl.* **1997**, *33*, 925–934. [CrossRef]
16. Wang, C.; Nasar, S.A.; Boldea, I. Vector control of three-phase flux reversal machine. *Electr. Mach. Power Syst.* **1999**, *28*, 153–166.
17. Wang, C.; Nasar, S.A.; Boldea, I. Three phase flux reversal machine (FRM). *IEE Proc. Electr. Power Appl.* **1999**, *146*, 139–146. [CrossRef]
18. Zhao, W.Y.; Chen, Y.G.; Shen, Y.B.; Xing, S.Z. Effective methods of reducing cogging torque in flux-reversal machine. *J. Iron Steel Res. Int.* **2006**, *13*, 444–449. [CrossRef]
19. Zhu, X.; Hua, W. An improved configuration for cogging torque reduction in flux-reversal permanent magnet machines. In Proceedings of the 2016 IEEE Conference on Electromagnetic Field Computation (CEFC), Miami, FL, USA, 13–16 November 2016; pp. 18–21.
20. Xiaofeng, Z.; Wei, H.; Zhongze, W. Cogging torque suppression in flux-reversal permanent magnet machines. *IET Electr. Power Appl.* **2018**, *12*, 135–143.
21. Zhang, B.; Wang, X.; Zhang, R.; Mou, X. Cogging torque reduction by combining teeth notching and rotor magnets skewing in PM BLDC with concentrated windings. In Proceedings of the 2008 International Conference on Electrical Machines and Systems, Wuhan, China, 17–20 October 2008; pp. 3189–3192.

22. Chu, W.Q.; Zhu, Z.Q. On-load cogging torque calculation in permanent magnet machines. *IEEE Trans. Magn.* **2013**, *49*, 2982–2989. [CrossRef]
23. Boldea, I.; Zhang, L.; Nasar, A. Theoretical characterization of flux reversal machines in low-speed servo drives- the pole-PM configuration. *IEEE Trans. Ind. Appl.* **2002**, *38*, 1549–1557. [CrossRef]
24. More, D.S.; Kalluru, H.; Fernandes, B.G. Outer rotor flux reversal permanent magnet machine for rooftop wind generator. In Proceedings of the Industry Applications Society Annual Meeting 2008 IAS' 08, Edmonton, AB, Canada, 25 April 2008; pp. 1–6.
25. More, D.S.; Fernandes, B.G. Power density improvement of three phase flux reversal with distributed winding. *IET Electr. Power Appl.* **2010**, *4*, 109–120. [CrossRef]
26. More, D.S.; Fernandes, B.G. Novel three phase flux reversal machine with full pitch winding. In Proceedings of the 7th International Conference on Power Electronics, Daegu, Korea, 22–26 October 2007; pp. 1007–1012.
27. Yang, H.; Zhu, Z.Q.; Lin, H.; Li, H.Y.; Lyu, S. Analysis of consequent-pole flux reversal permanent magnet machine with biased flux modulation theory. *IEEE Trans. Ind. Electron.* **2020**, *67*, 2107–2121. [CrossRef]
28. Luo, J.; Kou, B.; Zhang, H.; Qu, R. Development of a consequent pole transverse flux permanent magnet linear machine with passive secondary structure. *CES Trans. Electr. Mach. Syst.* **2019**, *3*, 39–44. [CrossRef]
29. Zhu, X.; Hua, W. Back-EMF waveform optimization of flux-reversal permanent magnet machines. *AIP Adv.* **2017**, *7*, 1–5. [CrossRef]
30. Dmitrievskii, V.; Prakht, V.; Sarapulov, S.; Askerov, D. A multipole single-phase SMC flux reversal motor for fans. In Proceedings of the 2016 XXII International Conference on Electrical Machines (ICEM), Lausanne, Switzerland, 4–7 September 2016; pp. 53–59.
31. Dmitrievskii, V.; Prakht, V.; Pozdeev, A.; Klimarev, V.; Mikhalitsyn, A. Angular grinder with new flux reversal motor. In Proceedings of the 2015 International Conference on Electrical Machines and Systems (ICEMS), Pattaya, Thailand, 25–28 October 2015; pp. 1366–1371.
32. Dmitrievskii, V.; Prakht, V. Gearless generator with magnets on the stator for wind turbine. In Proceedings of the Wind Europe Conference 2018, Hamburg, Germany, 25–28 September 2018; pp. 1–8.
33. Vidhya, B.; Srinivas, K.N. Effect of stator PM thickness and rotor geometry modifications on the minimization of cogging torque of flux reversal machine. *Turk. J. Electr. Eng. Comput. Sci.* **2017**, *25*, 4907–4922. [CrossRef]
34. Heoung, T.K.; Sung, H.W.; Bong, K.; Lee, J. Reduction of cogging torque in flux reversal machine by rotor teeth pairing. In Proceedings of the IEEE International Magnetics Conference (INTERMAG), Nagoya, Japan, 4–8 April 2005; pp. 3964–3966.
35. Gyeong, C.L.; Jung, T.U. Cogging torque reduction design of dual stator radial flux permanent magnet generator for small wing turbine. In Proceedings of the IEEE 2013 Tencon-Spring, Sydney, NSW, Australia, 17–19 April 2013; pp. 85–89.
36. Xu, W.; Zhu, J.; Zhang, Y.; Hu, J. Cogging torque reduction for radially laminated flux-switching permanent magnet machine with 12/14 poles. In Proceedings of the IECON 2011-37th Annual conference on IEEE Industrial Electronics Society, Melbourne, VIC, Australia, 7–10 November 2011; pp. 3590–3595.
37. Yusivar, F.; Roy, V.H.S.; Gunawan, R.; Halim, A. Cogging torque reduction with pole slot combination and notch. In Proceedings of the IEEE International Conference on Electrical Engineering and Computer Science, Kuta, Indonesia, 24–25 November 2014; pp. 260–263.

Publisher's Note: MDPI stays neutral with regard to jurisdictional claims in published maps and institutional affiliations.

energies

MDPI

Article

A Study on the Rotor Design of Line Start Synchronous Reluctance Motor for IE4 Efficiency and Improving Power Factor

Hyunwoo Kim [1], Yeji Park [1], Seung-Taek Oh [1], Hyungkwan Jang [1], Sung-Hong Won [2], Yon-Do Chun [3] and Ju Lee [1,*]

1 Department of Electrical Engineering, Hanyang University, Seoul 04763, Korea; khw7481@hanyang.ac.kr (H.K.); yejipark@hanyang.ac.kr (Y.P.); ost9107@naver.com (S.-T.O.); jhkinhyu@hanyang.ac.kr (H.J.)

2 Department of Electrical Engineering, Dongyang Mirae University, Seoul 08221, Korea; sagewide@dongyang.ac.kr

3 Electric Machines and Drives Research Center, Korea Electrotechnology Research Institute, Changwon 51543, Korea; ydchun@keri.re.kr

* Correspondence: julee@hanyang.ac.kr; Tel.: +82-2220-0342

Received: 15 October 2020; Accepted: 3 November 2020; Published: 4 November 2020

Abstract: As international regulations of motor efficiency are strengthened, the line-start synchronous reluctance motor (LS-SynRM) is being studied to improve the efficiency of the electrical motor in industrial applications. However, in industrial applications, the power factor is also an important performance index, but the LS-SynRM has poor power factor due to the saliency characteristic. In this paper, the rotor design of LS-SynRM is performed to improve the efficiency and power factor. First, the barrier design is performed to improve the efficiency and power factor using the response surface method (RSM). Second, the rotor slot design is performed according to the length of bar for synchronization. Lastly, the rib design is performed to satisfy the power factor and the mechanical reliability. The final model through the design process is analyzed using finite element analysis (FEA), and the objective performance is satisfied. To verify the FEA result, the final model is manufactured, and experiment is performed.

Keywords: finite element analysis; international electrotechnical committee; line start synchronous reluctance motor; power factor; super premium efficiency

1. Introduction

In industrial applications, electrical energy consumption of motors account for 35% to 40% of electrical energy generated in the world. If this electrical energy consumption can be reduced, several environmental impacts, such as the emissions of CO_2 or global warming, can be reduced [1,2]. For this reason, international regulations regarding motor efficiency were enacted by the International Electrotechnical Committee (IEC) 60034-30. According to IEC 60034-30, motor efficiency was classified as IE1 to IE5. Recently, the efficiency standards of industrial applications have been strengthened, and industrial motors may need to meet IE4 or IE5 efficiency class [3].

With various types of motors, the three-phase squirrel-cage induction motor (SCIM) is the most used, because of its simple structure, line-start ability, robustness, and low manufacturing cost. However, it is evident that the induction motor (IM) does not meet high efficiency due to rotor copper loss [4–6]. Due to a global trend on improving the efficiency of electric motors, there is ongoing research that focuses on this area. In References [7,8], the aluminum in the rotor slot is replaced by copper to reduce the rotor copper loss. Copper die-casting induction motors can improve efficiency by about 2–3% compared to the aluminum die-casting induction motors. However, copper has a higher melting point than aluminum, so a die-casting process is difficult, and the manufacturing cost increases. In industrial variable-speed applications, a variable-speed drive (VSD) is used to improve the efficiency of motors [9–11]. When using the VSD system, a synchronous motor can be controlled to improve the efficiency. However, the synchronous motor along with the VSD system is more expensive than SCIMs.

Line-start synchronous machines (LSSMs) are actively studied because they can meet the efficiency of IE4. In addition, a LSSM is direct-on-line (DOL) motor, such as IM, so it does not require an inverter, in which the cost of electric system can be reduced [12–17]. LSSM can be classified into two types; line-start permanent magnet synchronous motor (LSPM) and line-start synchronous reluctance motor (LS-SynRM). LSPM has the high efficiency and power factor because of the permanent magnet. However, the permanent magnet of LSPM must use a rare-earth magnet, such as neodymium, because of the high starting current. Furthermore, the resources of rare-earth magnets are limited, and lead to high costs [18,19]. Therefore, the LS-SynRM has received attention for its high efficiency and for the fact that it does not use a rare-earth magnet.

A reactive power of an electrical motor depends on the power factor. If the reactive power is increased by the low power factor, the copper loss in stator winding increases under the same output power and input voltage condition because of the current increases [20]. This increased copper loss decreases the efficiency of the electrical motor. For this reason, there are the power factor standards of industrial motors, according to IEC 60034. The electrical motor must meet the power factor standards to use industrial applications [21]. However, LS-SynRM has a lower power factor than IM and LSPM because of a saliency characteristic and may not meet the power factor standards [22–24]. Therefore, the design method considering the power factor is required.

This paper is a design process of LS-SynRM, for super premium efficiency, and for improving the power factor. The efficiency and the power factor of LS-SynRM is determined by *dq*-axis inductance, which is determined by the barrier design. Therefore, the design parameters are selected for IE4 efficiency and the power factor. Based on the selected design parameters, the barrier optimal design is performed using the response surface method (RSM) and finite element analysis (FEA). In addition, the rotor slot design is performed for synchronization and the rib design is performed to improve the power factor and satisfy the mechanical reliability through the safety factor. To verify the efficiency and the power factor of the design result, the final model is manufactured, and the experiment is performed to verify the IE4 efficiency and power factor.

This paper is organized as follows. In Section 2, the characteristics of LS-SynRM are discussed in respect to the structure and the operation in asynchronous and synchronous speed. In Section 3, the characteristic of the reference machine is analyzed. In Section 4, the design of LS-SynRM is performed using RSM and FEA and the final model is analyzed using FEA. In Section 5, to verify FEA results, the experiment is performed and the results of LS-SynRM are compared with FEA results. Finally, Section 6 contains the conclusion.

2. Characteristic of LS-SynRM

2.1. Structure of LS-SynRM

Figure 1 shows the rotor structure of a general LS-SynRM. LS-SynRM has a structure in which the squirrel cage slots are placed on the outer side of the rotor, and the barrier is arranged on the inner side of the rotor. Because of this structure, LS-SynRM operates as IMs at the asynchronous speed and as SynRM at the synchronous speed. Moreover, the rotor design of LS-SynRM is classified as a rotor slot and barrier. The rotor slot design determines a starting characteristic of LS-SynRM and the barrier design determines the performance of LS-SynRM, such as the efficiency and power factor. Therefore, the rotor slot design is important to reach the synchronous speed, and the barrier design is important to satisfy the super-premium efficiency (IE4).

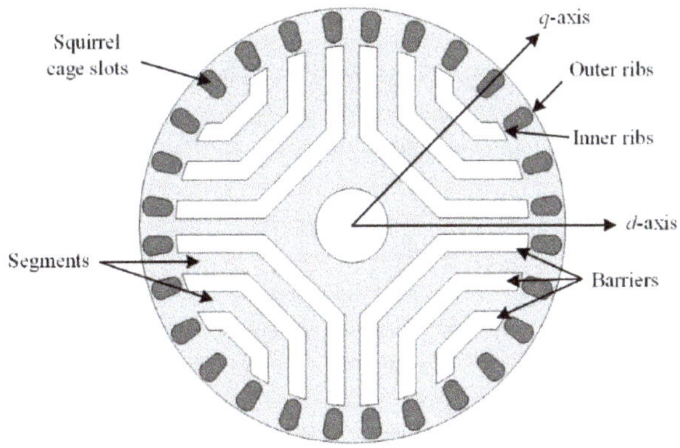

Figure 1. Rotor structure of the line-start synchronous reluctance motor (LS-SynRM).

2.2. Asynchronous Speed

In asynchronous speed, the LS-SynRM has the slip, which generates an induced voltage on a conductor in a rotor slot. An induced current that is generated by the induced voltage provides the magnetic torque that LS-SynRM can be synchronization. Moreover, in case of the LS-SynRM, the reluctance torque is generated due to a saliency difference, which is the difference between *dq*-axis inductance. Therefore, the torque of LS-SynRM is expressed as Equation (1) [25,26].

$$T_e = T_{cage} + T_{rel}\sin(2s\omega t + \alpha) \tag{1}$$

where T_e is the torque of LS-SynRM at asynchronous speed, T_{cage} is the magnetic torque generated by the induced current, T_{rel} is the amplitude of reluctance torque produced by the saliency, s is slip, ω is the electrical angular frequency, and α is the phase angle of pulsating torque.

The magnetic torque is constant and depends on the slip as in the induction motor, and the average of reluctance torque is zero, but this torque pulsates twice the slip frequency. Figure 2 shows a speed-torque curve of LS-SynRM at an asynchronous speed. The amplitude of the pulsating torque and the average torque depend on the slip. In a synchronous speed, the maximum torque is called the pull-out torque, which is the maximum load torque that can be synchronized [26].

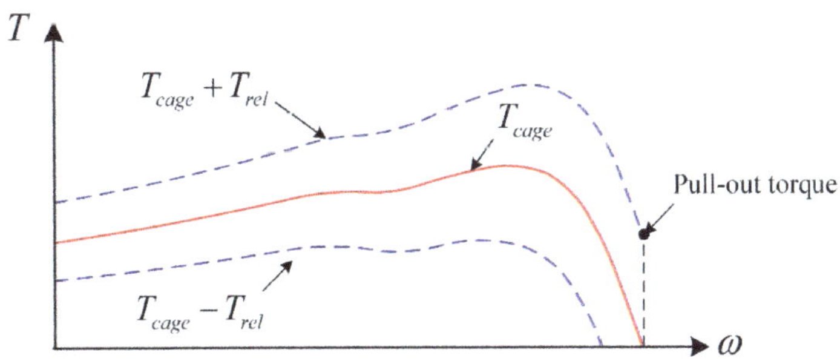

Figure 2. Speed-torque curve of LS-SynRM.

2.3. Synchronous Speed

At a synchronous speed, there are no eddy currents in the rotor slot because the slip is zero. Therefore, the LS-SynRM is operated as the synchronous reluctance motor and the efficiency is improved compared to SCIM due to the secondary copper loss. The efficiency of LS-SynRM is expressed as follows.

$$\eta = \frac{P_{out}}{P_{in}} = \frac{P_{out}}{P_{out}+P_{loss}} = \frac{T_e\omega_e}{T_e\omega_e+P_{copper}+P_{core}}$$
$$T_e = \frac{3}{2}\frac{P}{2}(L_d - L_q)I_a^2 \sin 2\beta \quad (2)$$

where η is the efficiency, P_{out} is mechanical power, P_{in} is electrical input power, P_{loss} is total loss of motor, T_e is torque of motor, ω_e is synchronous speed, P_{copper} is copper loss, and P_{core} is core loss, P is the number of pole, L_d and L_q are d-, q-axis inductance, respectively, I_a is the current, and β is the current phase angle.

The efficiency is determined by the reluctance torque, the core loss, and the copper loss. The core loss depends on the magnetic flux density and input frequency, which is determined when designing electric machines. Therefore, the copper loss must be reduced to improve the efficiency. This should increase the torque per current, which should increase the rotor saliency difference.

The power factor of the LS-SynRM is determined by the phase difference between the voltage and current. Figure 3a shows a vector diagram of the LS-SynRM, and the power factor of LS-SynRM can be expressed as follows.

$$\cos\varphi = \frac{\rho - 1}{\sqrt{\frac{\rho^2}{\cos^2\beta} + \frac{1}{\sin^2\beta}}}, \quad \rho = \frac{L_d}{L_q} \quad (3)$$

where cosφ is the power factor, and ρ is the saliency ratio.

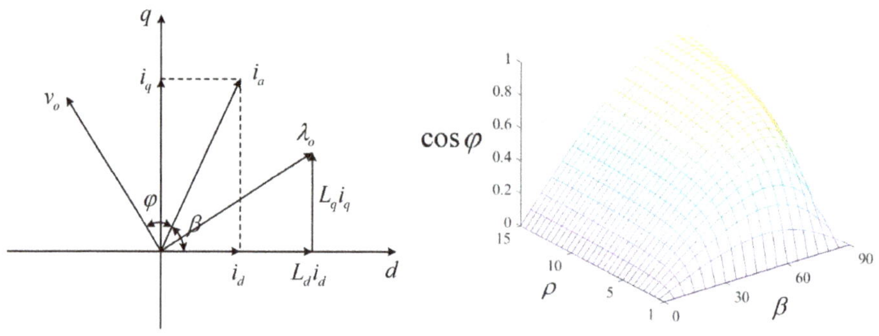

(**a**) (**b**)

Figure 3. Characteristic of LS-SynRM (**a**) vector diagram, and (**b**) power factor.

The power factor is determined by the saliency ratio and the current phase angle. Figure 3b shows the power factor according to the saliency ratio and the current phase angle. The power factor increases as the saliency ratio increases. Therefore, the saliency difference and ratio are important design parameters for IE4 efficiency and improving the power factor.

3. Reference Machine

In order to compare the efficiency and power factor of LS-SynRM, 2.2 kW four pole IM is selected as the reference motor. Figure 4 show the FEA model and Table 1 shows the specification of reference machine. ANSYS Maxwell was used to analyze the performance of the reference machine.

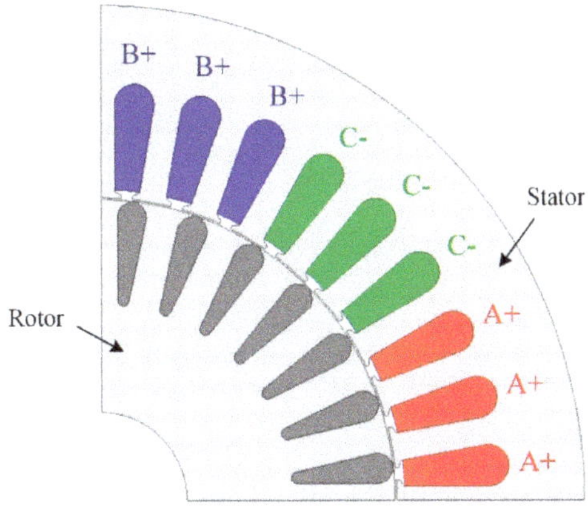

Figure 4. Finite element analysis (FEA) model of reference machine.

Table 1. Specification of reference machine.

Item	Value	Unit
Output power	2.2	kW
Line voltage	380	V_{rms}
Input frequency	60	Hz
Outer diameter (stator/rotor)	180/109.4	mm
Inner diameter (stator/rotor)	110/32	mm
Number of slot (stator/rotor)	36/28	-
Stack length	120	mm
Air gap length	0.3	mm

Table 2 shows the FEA and experiment results of the reference machine. The efficiency of the reference machine is 90.7% that is IE3 efficiency according to IEC 60034. Because IE4 efficiency is 91%, the reference machine does not satisfy IE4 efficiency. In general, the induction motor has the rotor copper loss due to the slip, and this loss accounts for 23% of total losses in Table 2. If the rotor copper loss is reduced, the efficiency of the electrical machine can satisfy IE4 efficiency. LS-SynRM does not have the rotor copper loss because this motor is operated at a synchronous speed, where slip is zero. Therefore, LS-SynRM can satisfy IE4 efficiency, and the design process of LS-SynRM has studied to improve the efficiency.

Table 2. FEA and experiment result of reference model.

Item	Value		Unit
	FEA	Experiment	
Power	2.2	2.2	kW
Speed	1764.96	1762	rpm
Torque	12	11.9	Nm
Core loss	58.27	63	W
Stator copper loss	72.6	78	W
Rotor copper loss	40.53	51	W
Total loss	214.3	220	W
Efficiency	91.2	90.7	%
Power factor	83.8	80	%

4. Design of LS-SynRM

4.1. Design Process

Figure 5 shows the design process of LS-SynRM for IE4 efficiency and improving the power factor. The stator, rotor, and number of rotor slot are constrained under the same conditions as the reference machine. The main design process is three steps. First, the design of the barrier is performed to improve the efficiency and power factor using RSM. The efficiency and power factor of LS-SynRM are determined by the saliency difference and ratio. Therefore, there are two steps in the design of the barriers: the number of barriers and thickness of barriers and segments. Second, the design of the rotor slot is performed to reach LS-SynRM into the synchronous speed. Because synchronization is determined by the resistance and leakage inductance of the rotor bar, the design is performed according to the depth of the rotor slot. Lastly, the design of each rib is performed considering the power factor and mechanical reliability. The thickness of the ribs affects the leakage flux, which affect the performance of LS-SynRM, and is determined by the thickness of each rib. Therefore, the thickness of each rib must be designed considering the performance and safety factor. If the efficiency and power factor does not satisfy, the design parameter and range are reselected, and RSM is performed to satisfy the performance. Table 3 shows IEC 60034 standard and the design objective of the efficiency and power factor. Considering the margin based on Table 2, FEM performance is selected as 91.5% efficiency and 81% power factor.

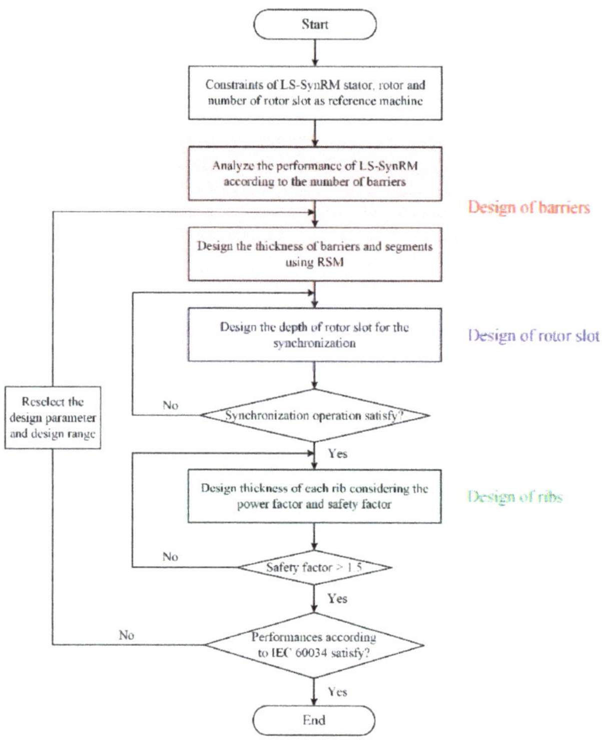

Figure 5. Design process of LS-SynRM.

Table 3. International Electrotechnical Committee (IEC) 60034 standard and design objective of the efficiency and power factor.

Items		IEC 60034	Design Objective	Unit
Performance	Efficiency	91	≥91.5	%
	Power factor	77	≥81	%

4.2. *Design of Barrier*

From Equation (3), the power factor of LS-SynRM is determined by the saliency ratio, which is ratio *d*-axis inductance to *q*-axis inductance. The *dq*-axis inductances are determined by the number of barriers and thickness of flux barrier and segments [27]. Therefore, the barrier design is performed according to the number of barriers and thickness of flux barrier and segments.

4.2.1. Number of Barrier

Figure 6 shows the FEA model according to the number of barriers. The number of rotor slots is the same as the reference machine, and the total length of barrier and segment is constant, for comparison under the same conditions. Furthermore, for each layer of each model, the thickness of the barrier and segment are compared equally. Table 4 shows the comparison of the *dq*-axis inductances, the saliency ratio, the power factor, and efficiency under the rated current condition. From Table 4, the larger the saliency ratio, the larger the power factor. Moreover, the efficiency of each model satisfies more than IE4 efficiency. Therefore, the design of the barrier, according to the thickness of the flux barrier and segments, is performed based on model 3.

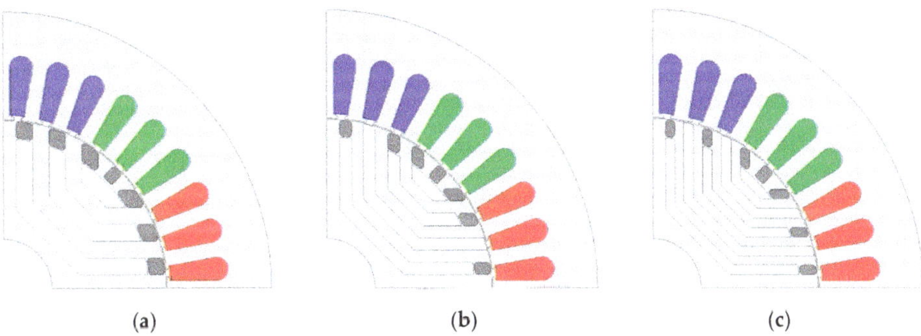

Figure 6. FEA model according to the number of barriers (**a**) model 1, (**b**) model 2, and (**c**) model 3.

Table 4. FEA result of LS-SynRM according to the number of barriers.

Model	L_d (mH)	L_q (mH)	Saliency Ratio	Power Factor (%)
Model 1	33.6	301.6	8.97	78.7
Model 2	31.7	297.7	9.38	79.3
Model 3	31.2	309.5	9.92	80.5

4.2.2. Thickness of Flux Barriers and Segments

To maximize the power factor, the optimal design is performed using RSM and FEA. Figure 7 shows the design parameters for the optimal design [28]. The range of design parameters is selected within the rotor constraints, as shown in Table 5. Furthermore, the objective function is maximizing the power factor and the efficiency. Figure 8a shows the RSM result and Figure 8b shows the optimal design result. Table 6 shows the optimal design result. In Table 6, RSM and FEA results are similar, so the optimal result is valid.

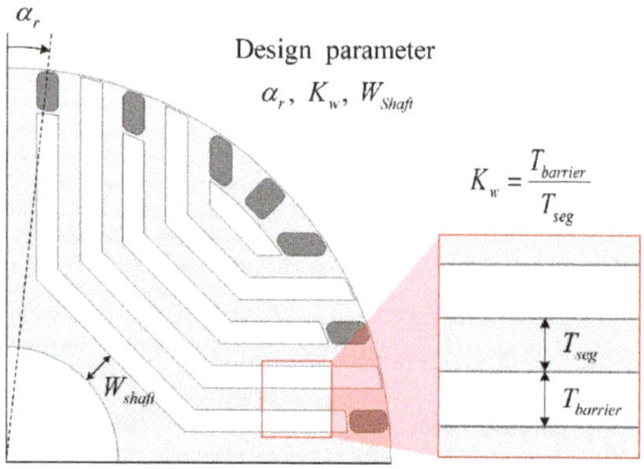

Figure 7. Design parameters for the optimal design using the response surface method (RSM).

Table 5. Range of design parameters.

Items	α_r (degM)	K_w	W_{shaft} (mm)
Range	4.5–9	0.4–0.6	3–6

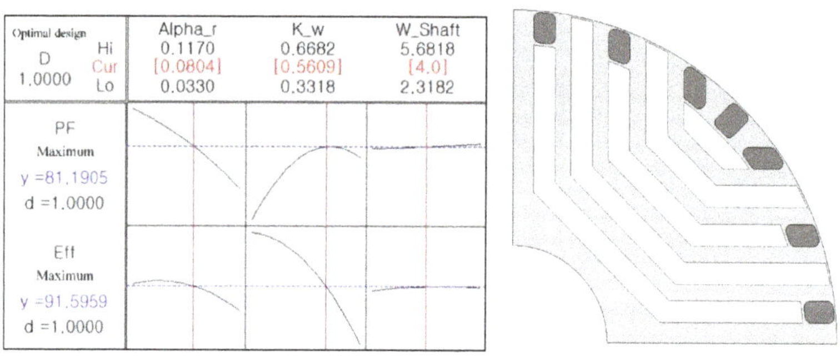

(a) (b)

Figure 8. Optimization using RSM under rated current condition. (**a**) RSM result, (**b**) optimal design result.

Table 6. Optimal design result.

Parameter			Power Factor (%)		Efficiency (%)	
α_r (degM)	K_w	W_{shaft} (mm)	RSM	FEA	RSM	FEA
7.236	0.5609	4	81.19	81.02	91.59	91.75

4.3. Design of Rotor Slot

In general, the starting characteristic is determined by the rotor resistance and leakage inductance. This rotor resistance and leakage is determined by the depth and thickness of the rotor slot. However, the barriers are constraints on the thickness and depth of the rotor slot. Therefore, for synchronization, the design parameters of the rotor slot are determined, as shown in Figure 9a. In order to analyze synchronization, the analysis, according the selected design parameters, is performed by the time-step FEA, which is the mechanical and electromagnetic transient analysis [29,30], considering inertia of LS-SynRM. Figure 9b shows the time step FEA result. The synchronization region is determined by the depth of the rotor slot. Figure 10 shows the time step FEA result for the three selected points. If synchronization fails, the torque and speed pulsate. As a result, the design parameter is determined as T_{bar1} = 20 (mm) and T_{bar2} = 13 (mm).

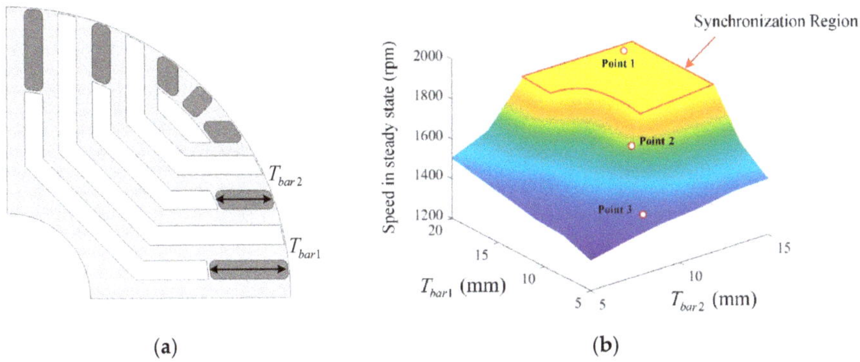

Figure 9. Design parameter of rotor slot and FEA result (**a**) design parameter, (**b**) FEA result.

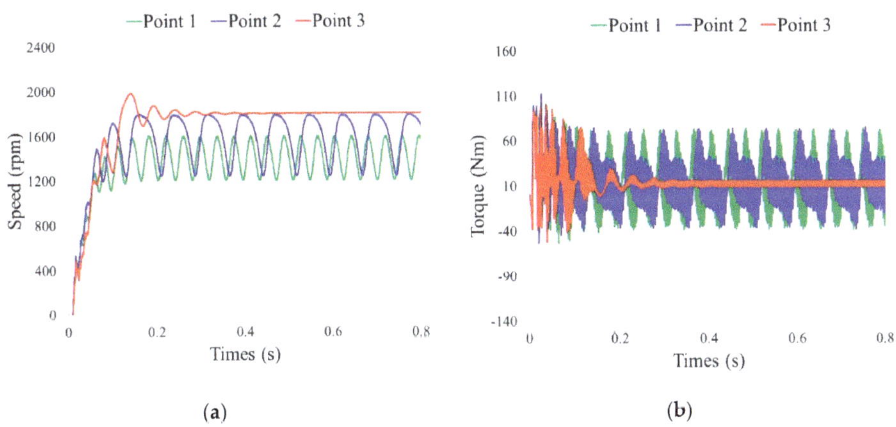

Figure 10. Speed and torque curve of the selected three points (**a**) speed curve (**b**) torque curve.

4.4. Design of Ribs

When LS-SynRM rotates, the centrifugal force is focused on the outer rib. If the thickness of the outer rib is designed to be small, the LS-SynRM is mechanically damaged during operation. Therefore, the design of rib must be performed for use as the industrial application. However, the thickness of the outer rib also affects the power factor, so the design of the rib is performed considering the power factor and mechanical reliability. The mechanical reliability is determined by a safety factor, as shown in the following equation. Generally, the mechanical reliability is ensured if the safety factor is higher than 1.5 [31].

$$SF = \frac{\sigma_{\text{yield}}}{\sigma_{\text{max}}} \tag{4}$$

where SF is the safety factor, σ_{yield} is the tensile yield strength of material, σ_{max} is the stress of material during rotation.

Because of the structure of LS-SynRM, there are several bridges that reduce the mechanical stress at the rib. Therefore, the design parameters are selected, as shown in Figure 11. Considering the manufacture, the design parameters are designed to be 0.3 (mm) or more. Figure 12 shows the power factor and the safety factor, according to the design parameters. As the thickness of each rib increases, the power factor is reduced because the leakage magnetic flux is increased. Furthermore, the safety factor is increased due to reduced mechanical stress. Considering the objective performance in Table 3, the design parameters are designed as T_{rib1} = 0.3 (mm) and T_{rib2} = 0.7 (mm).

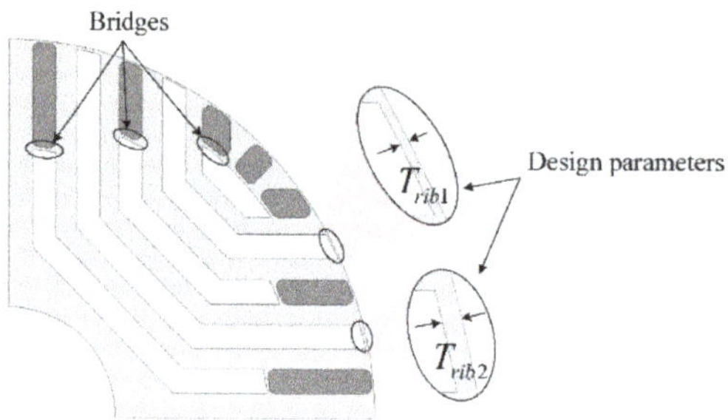

Figure 11. Design parameter for the power factor and safety factor.

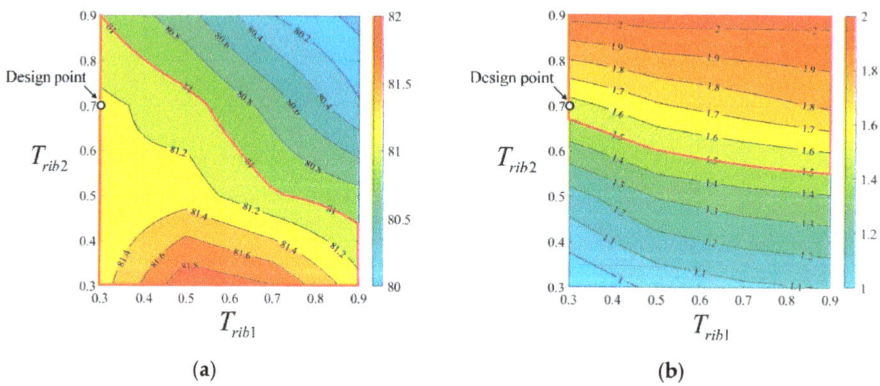

Figure 12. FEA result according to the design parameter of rib (**a**) safety factor, and (**b**) power factor.

4.5. Design Result

Figure 13 shows the design result through the design process of LS-SynRM. Table 7 shows the FEA result of LS-SynRM. The efficiency of LS-SynRM is 91.7% and the power factor of LS-SynRM is 81.2%. Compared with Table 2, the efficiency is improved about 0.5%, but the power factor is decreased about 2.6% than the reference machine. The efficiency satisfies the IE4 efficiency and the power factor satisfies the IEC 60034 standard for the industrial application. Therefore, the LS-SynRM can improve the efficiency compared with IM and be used as an industrial electric machine.

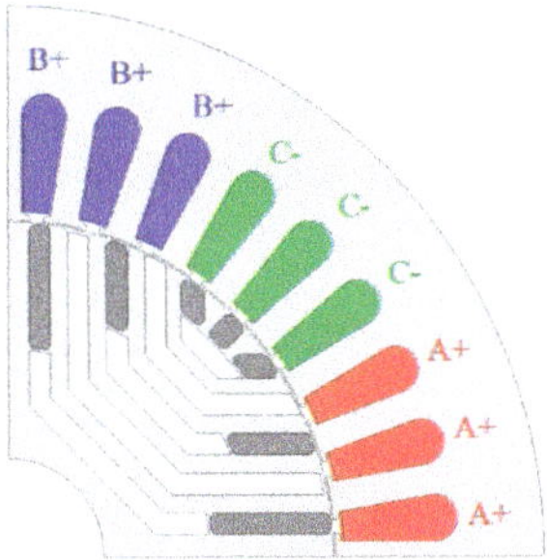

Figure 13. Final model of LS-SynRM.

Table 7. FEA result of the final LS-SynRM.

Item	Value	Unit
Power	2.2	kW
Speed	1800	rpm
Torque	11.7	Nm
Core loss	53.3	W
Stator copper loss	87.78	W
Rotor copper loss	12.3	W
Total loss	201.4	W
Efficiency	91.7	%
Power factor	81.2	%
Safety factor	1.56	-

5. Experimental Validation

5.1. Manufacture of LS-SynRM

To verify the FEA result, the designed LS-SynRM is manufactured. Figure 14a shows the rotor core of the manufactured LS-SynRM. A die-casting process is performed to inject the melted aluminum into the rotor slot. During the die-casting process, a high press is applied to the rotor slot so that the melted aluminum is injected in the rotor slot. Figure 14b shows the endplate, which is required to prevent the melted aluminum from being injected into the barrier. Figure 14c shows the cross-section of the rotor and the aluminum is filled in the rotor slot. Figure 14d shows the rotor with the end-ring. The width of the end-ring is the same as the reference machine and the height of the end-ring is the difference between the deepest rotor slot depth and the outer diameter of the rotor.

5.2. Experiment Result

The experiment is performed to verify the FEA result. Figure 15 shows the experiment environment. The dynamometer motor is induction motor as 2.2 kW load. Using voltage and frequency control, LS-SynRM runs up and reaches a synchronous speed. After the LS-SynRM reaches a synchronous speed, the load torque is applied through the dynamometer motor to measure the efficiency and power factor of LS-SynRM. In the power analyzer, the efficiency and power factor are calculated using measured voltage, current, torque, and speed. Table 8 shows the comparison of FEA and the experiment result of LS-SynRM. When comparing the experiment and FEA results, the experiment current increases by about 5% than the FEA current. This reduces the power factor and increases the copper loss, as shown in Table 8. The power factor of FEA and experiment is 81.2% and 77.3%, respectively. However, the efficiencies of FEA and the experiment are similar because the total losses are similar. As a result, it is confirmed that the IE4 class efficiency and the power factor in IEC 60034 is satisfied.

Figure 14. Manufactured LS-SynRM (**a**) rotor core, (**b**) endplate for die-casting, (**c**) cross-section of rotor, (**d**) rotor with the end-ring.

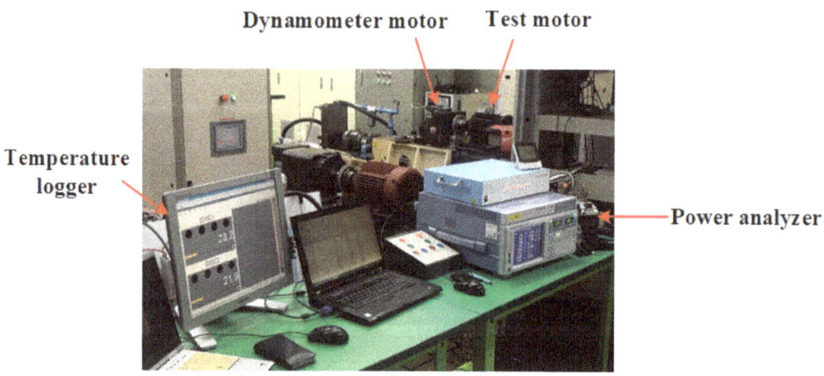

Figure 15. The experiment environment and test dynamometer.

Table 8. Comparison of FEA and experiment result.

Item	Value		Unit
	FEA	Experiment	
Power	2.2	2.2	kW
Torque	11.7	11.6	Nm
Current	4.56	4.8	A_{rms}
Core loss	53.3	45.9	W
Stator copper loss	87.78	100	W
Rotor copper loss	12.3	0	W
Total loss	201.4	201.7	W
Efficiency	91.7	91.6	%
Power factor	81.2	77.3	%

6. Conclusions

In this paper, the design method of LS-SynRM is studied to satisfy the efficiency and power factor of IEC 60034 standard. In addition, LS-SynRM is designed considering the mechanical reliability for use in the industrial application. The performance of LS-SynRM is analyzed according to several design parameters, such as the number of barriers, the thickness of the barriers, and the segments, the depth of the rotor slot, and thickness of each rib. The design of the barrier is performed to satisfy the performance of LS-SynRM using RSM. The design of the rotor slot is performed, and the synchronization region is determined by the depth of the rotor slot. The design of each rib is performed, and the thickness of each rib is designed considering the power factor and the safety factor. Considering the efficiency, the power factor, and the safety factor, the optimal LS-SynRM is designed. The final LS-SynRM satisfies the efficiency of 91.7%, the power factor of 81.2%, and the safety factor 1.56. The final LS-SynRM is manufactured and the experiment is performed to verify the FEA result. As a result, the efficiency and power factor satisfy the IEC 60034 standard.

Author Contributions: Conceptualization, H.K.; methodology, H.J.; validation, S.-T.O.; formal analysis, H.K. and S.-T.O.; investigation, H.K. and Y.P.; resources, S.-H.W.; data curation, Y.P.; writing—original draft preparation, H.K.; writing—review and editing, S.-T.O. and H.J.; supervision, S.-H.W. and J.L.; project administration, J.L.; funding acquisition, Y.-D.C. and S.-H.W. All authors have read and agreed to the published version of the manuscript.

Funding: This research was funded by the Korea Agency for Infrastructure Technology Advancement (KAIA) grant funded by the Korea government Ministry of Ministry of Land, Infrastructure and Transport (20CTAP-C157784-01) and in part by a Korea Institute of Energy Technology Evaluation and Planning grant funded by the Korea government (20192010106780, A Construction and Operation of Open Platform for Next-Generation Super Premium Efficiency (IE4) Motors).

Conflicts of Interest: The authors declare no conflict of interest.

References

1. Almeida, A.T.D.; Ferreira, F.J.T.E.; Baoming, G. Beyond Induction Motors-Technology Trends to Move Up Efficiency. *IEEE Ind. Appl.* **2014**, *50*, 2103–2114.
2. Almeida, A.T.D.; Ferreira, F.J.T.E.; Duarte, A.Q. Technical and Economical Considerations on Super High-Efficiency Three-Phase Motors. *IEEE Ind. Appl.* **2014**, *50*, 1274–1285.
3. Ferreira, F.J.T.E.; Baoming, G.; Almeida, A.T.D. Reliability and Operation of High-Efficiency Induction Motors. *IEEE Ind. Appl.* **2016**, *52*, 4628–4637.
4. Rafajdus, P.; Hrabovcova, V.; Lehocky, P.; Makys, P.; Holub, F. Effect of Saturation on Field Oriented Control of the New Designed Reluctance Synchronous Motor. *Energies* **2018**, *11*, 3223. [CrossRef]
5. Aguba, V.; Muteba, M.; Nicolae, D.V. Transient Analysis of a Start-up Synchronous Reluctance Motor with Symmetrical Distributed Rotor Cage Bars. In Proceedings of the 2017 IEEE AFRICON, Cape Town, South Africa, 18–20 September 2017.
6. Xie, Y.; Pi, C.; Li, Z. Study on Design and Vibration Reduction Optimization of High Starting Torque Induction Motor. *Energies* **2019**, *12*, 1263. [CrossRef]

7. Kim, D.-J.; Hong, D.-K.; Choi, J.-H.; Chun, Y.-D.; Woo, B.-C.; Koo, D.-H. An Analytical Approach for a High Speed and High Efficiency Induction Motor Considering Magnetic and Mechanical Problems. *IEEE Trans. Magn.* **2013**, *49*, 2319–2322. [CrossRef]
8. Zhang, Q.; Liu, H.; Zhang, Z.; Song, T. A Cast Copper Rotor Induction Motor for Small Commercial EV Traction: Electromagnetic Design, Analysis, and Experimental Tests. *CES Trans. Electr. Mach. Syst.* **2018**, *2*, 417–424. [CrossRef]
9. Kazakbaev, V.; Prakht, V.; Dmitrievskii, V.; Ibrahim, M.N.; Oshurbekov, S.; Sarapulov, S. Efficiency Analysis of Low Electric Power Drives Employing Induction and Synchronous Reluctance Motors in Pump Applications. *Energies* **2019**, *12*, 1144. [CrossRef]
10. Aarniovuori, L.; Kolehmainen, J.; Kosonen, A.; Niemela, M.; Chen, H.; Cao, W.; Pyrhonen, J. Application of Calorimetric Method for Loss Measurement of a SynRM Drive System. *IEEE Trans. Ind. Electron.* **2016**, *64*, 2005–2015. [CrossRef]
11. Goetzler, W.; Sutherland, T.; Reis, C. *Energy Savings Potential and Opportunities for High Efficiency Electric Motors in Residential and Commercial Equipment*; United States Department of Energy: Washington, DC, USA, 2013.
12. Ghoroghchian, F.; Aliabad, A.D.; Amiri, E. Design improvement of dual-pole LSPM synchronous motor. *IET Electr. Power Appl.* **2019**, *13*, 742–749. [CrossRef]
13. Ghahfarokhi, M.M.; Aliabad, A.D.; Boroujeni, S.T.; Amiri, E.; Faradonbeh, V.Z. Analytical modelling and optimization of line start LSPM synchronous motors. *IET Electr. Power Appl.* **2020**, *14*, 398–408.
14. Mingardi, D.; Bianchi, N. Line-Start PM-Assisted Synchronous Motor Design, Optimization, and Tests. *IEEE Trans. Ind. Electron.* **2017**, *64*, 9739–9747. [CrossRef]
15. Liu, H.C.; Lee, J. Optimum Design of an IE4 Line-Start Synchronous Reluctance Motor Considering Manufacturing Process Loss Effect. *IEEE Trans. Ind. Electron.* **2018**, *65*, 3104–3114. [CrossRef]
16. Tampio, J.; Kansakangas, T.; Suuriniemi, S.; Kolehmainen, J.; Kettunen, L.; Ikaheimo, J. Analysis of Direct-On-Line Synchronous Reluctance Machine Start-Up Using a Magnetic Field Decomposition. *IEEE Trans. Ind. Appl.* **2017**, *53*, 1852–1859.
17. Liu, H.-C.; Hong, H.-S.; Cho, S.; Lee, J.; Jin, C.-S. Bubbles and Blisters Impact on Diecasting Cage to the Designs and Operations of Line-Start Synchronous Reluctance Motors. *IEEE Trans. Magn.* **2017**, *53*, 8202504. [CrossRef]
18. Dent, P. Rare earth elements and permanent magnets. *J. Appl. Phys.* **2012**, *111*, 07A721. [CrossRef]
19. Goss, J.; Popescu, M.; Staton, D. A comparison of an interior permanent magnet and copper rotor induction motor in a hybrid electric vehicle application. In Proceedings of the International Electric Machines & Drives Conference, Chicago, IL, USA, 12–15 May 2013; pp. 220–225.
20. Kazakbaev, V.; Parkht, V.; Dmitrievskii, V.; Oshurbekov, S.; Golovanov, D. Life Cycle Energy Cost Assessment for Pump Units with Various Types of Line-Start Operating Motors Including Cable Losses. *Energies* **2020**, *13*, 3546–3559.
21. Kim, H.; Park, Y.; Liu, H.-C.; Han, P.-W.; Lee, J. Study on Line-Start Permanent Magnet Assistance Synchronous Reluctance Motor for Improving Efficiency and Power Factor. *Energies* **2020**, *13*, 384. [CrossRef]
22. Joo, K.-J.; Kim, I.-G.; Lee, J.; Go, S.-C. Robust Speed Sensorless Control to Estimated Error for PMa-SynRM. *IEEE Trans. Magn.* **2017**, *53*, 8102604.
23. Ozcelik, N.G.; Dogru, U.E.; Imeryuz, M.; Ergene, L.T. Synchronous Reluctance Motor vs. Induction Motor at Low-Power Industrial Applications: Design and Comparison. *Energies* **2019**, *12*, 2190. [CrossRef]
24. Ogunjuyigbe, A.S.O.; Jimoh, A.A.; Ncolae, D.V.; Obe, E.S. Analysis of synchronous reluctance machine with magnetically coupled three-phase windings and reactive power compensation. *IET Electr. Power Appl.* **2010**, *4*, 291–303. [CrossRef]
25. Kersten, A.; Liu, Y.; Pehrman, D.; Thiringer, T. Rotor Design of Line-Start Synchronous Reluctance Machine With Round Bars. *IEEE Ind. Appl.* **2019**, *55*, 3685–3696. [CrossRef]
26. Gamba, M.; Pellegrino, G.; Vagati, A.; Villata, F. Design of a Line-Start Synchronous Reluctance Motor. In Proceedings of the 2013 International Electric Machines & Drives Conference, Chicago, IL, USA, 12–15 May 2013.
27. Pellegrino, G.; Cupertino, F.; Gerada, C. Automatic Design of Synchronous Reluctance Motors Focusing on Barrier Shape Optimization. *IEEE Trans. Ind. Appl.* **2015**, *51*, 1465–1474. [CrossRef]
28. Kim, K.-C.; Ahn, J.S.; Won, S.H.; Hong, J.-P.; Lee, J. A Study on the Optimal Design of SynRM for the High Torque and Power Factor. *IEEE Trans. Magn.* **2007**, *43*, 2543–2545. [CrossRef]

29. Liu, H.C.; Seol, H.S.; Kim, J.Y.; Lee, J. Design and Analysis of an IE4 Class Line-Start Synchronous Reluctance Motor Considering Total Loss and Starting Performance. *J. Electron. Mater.* **2019**, *48*, 1386–1394. [CrossRef]
30. Wang, Y.; Chau, K.T.; Chan, C.C.; Jiang, J.Z. Transient Analysis of a New Outer-Rotor Permanent-Magnet Brushless DC Drive Using Circuit-Field-Torque Coupled Time-Stepping Finite-Element Method. *IEEE Trans. Magn.* **2002**, *38*, 1297–1300. [CrossRef]
31. Lee, J.-K.; Jung, D.-H.; Lim, J.; Lee, K.-D.; Lee, J. A Study on the Synchronous Reluctance Motor Design for High Torque by Using RSM. *IEEE Trans. Magn.* **2018**, *54*, 8103005. [CrossRef]

energies

MDPI

Article

Investigation of Torque Performance and Flux Reversal Reduction of a Three-Phase 12/8 Switched Reluctance Motor Based on Winding Arrangement

Ruchao Pupadubsin, Seubsuang Kachapornkul, Prapon Jitkreeyarn, Pakasit Somsiri and Kanokvate Tungpimolrut *

Machines and Power Conversions Research Team, National Electronics and Computer Technology Center, National Science Technology Development Agency, Pathum Thani 12120, Thailand; ruchao.pupadubsin@nectec.or.th (R.P.); seubsuang.kachapornkul@nectec.or.th (S.K.); prapon.jitkreeyarn@nectec.or.th (P.J.); pakasit.somsiri@nectec.or.th (P.S.)
* Correspondence: kanokvate.tungpimolrut@nectec.or.th

Abstract: The goal of this paper is to present a comparative analysis of two types of winding arrangements for a three-phase 12/8 switched reluctance motor (SRM), where short- and fully-pitched winding arrangements under unipolar operation are considered. From the analytical results, the short-pitched winding has the best torque per copper weight ratio. The core loss based on counting the number of flux reversals in the stator yoke for each winding arrangement is also proposed and mentioned. To reduce the magnetic flux reversals in the stator core, changing the direction of the magnetic flux path by modifying the winding polarities of the short-pitched winding could reduce 10–13% of core loss compared to the conventional winding. A 1 kW, 12/8 SRM prototype for the ventilation fan application is constructed and tested in order to verify the design consideration of winding configuration. At the rated condition, a maximum efficiency around 82.5% could be achieved.

Keywords: switched reluctance machine; winding arrangement; static torque; dynamic torque; torque ripple; core loss; flux reversal; leakage-flux; magnetic flux path

Citation: Pupadubsin, R.; Kachapornkul, S.; Jitkreeyarn, P.; Somsiri, P.; Tungpimolrut, K. Investigation of Torque Performance and Flux Reversal Reduction of a Three-Phase 12/8 Switched Reluctance Motor Based on Winding Arrangement. *Energies* **2022**, *15*, 284. https://doi.org/10.3390/en15010284

Academic Editors: Vladimir Prakht, Mohamed N. Ibrahim and Vadim Kazakbaev

Received: 29 November 2021
Accepted: 29 December 2021
Published: 1 January 2022

1. Introduction

The switched reluctance motor (SRM) [1,2] is a double salient machine having a simple magnetic structure. It has only concentrated windings on the stator and without any coil winding or permanent magnet on the rotor, resulting in cost savings for the manufacturing process. Other advantages of SRM include high starting torque, fault tolerant, high efficiency over wide speed range and broad choice of converters depending on the application requirements; these make SRM an attractive alternative to conventional drive in numerous applications.

Many modern applications require higher efficiency and higher performance machines with reducing production cost. The challenge issue of SRM design techniques is to increase torque density while minimizing torque ripple, as these are major disadvantages of SRMs when compared to permanent–magnet (PM) motors. Most of the design methods are directly implemented via the design of motor geometries such as stator/rotor pole arc, stator/rotor pole shape, air gap, and stack length. To reduce the manufacturing cost, research work in [3] introduced the optimization algorithm for SRM design to achieve a maximum torque at the minimum mass of the motor. The recent optimization design technique for finding the maximum torque density of 12/8 SRM was revealed in [4]. This algorithm, based on the Grey Wolf Optimizer (GWO), can increase torque by 120% at the same outer value as the original design.

Changing winding arrangements is a simple method that does not require the machine structure to be redesigned, and a substantial increase in efficiency or performance can be achieved. For example, with a fully-pitched winding arrangement, as reported in [5–7], torque capability can be increased by over 30%, resulting in higher efficiency than the conventional winding arrangement. The fully-pitched winding, which can operate with either unipolar or bipolar operation, uses the mutual-coupling among the phases for torque production and enhancement of torque density. In [8,9], an analysis, design, and SRM operation of the fractionally-pitched winding arrangement that also utilizes the mutual–coupling between phases are discussed. However, the fractionally-pitched winding is suitable for bipolar operation rather than unipolar operation. Another investigation on winding arrangements [10,11] uses finite element analysis to study short- and long-flux paths using single-phase and double-phase excitation for a three-phase 12/8 SRM; the studies also reveal that the maximum torque can be increased.

The flux reversal has a significant impact on the core loss of the SRM. Research on reducing flux reversal, such as [12–16], was implemented by modifying the shape of the stator. In this paper, the method to reduce the flux reversal will be discussed based on the winding arrangement.

The objective of this paper is to investigate the winding arrangements of a three-phase 12/8 SRM under unipolar operation by considering the torque density comparison based on the same conduction loss, including the core loss effects, the total winding weights, and torque ripple of different winding arrangements. The simple method, which is the flux reversal pattern, for evaluating the core loss of the stator in each winding arrangement is introduced. Furthermore, the winding selection procedure for a three-phase 12/8 SRM to achieve high torque density, low torque ripple, and cost-saving for manufacturing is described. The finite element program (JMAG) and PC–SRD software are used mainly for the calculation of the torque density (static and dynamic), core loss, and structure of winding; winding weight, slot–fill factor, etc.

2. Winding Arrangement

2.1. Short-Pitched Winding

Figure 1 shows three diagrams of the short-pitched winding configuration and flux paths of the three-phase 12/8 SRM when a single phase (phase A) is excited with unipolar operation.

For the short-pitched winding, each coil is wound around a single stator pole. The winding's polarities are arranged in the stator to give a magnetic pole when each phase is excited, resulting in the stator poles acting as a north (N) or south (S) magnetic pole, which is the main cause of magnetic–flux path direction. The 12/8 SRM's winding polarities of short-pitched winding are modified to change the magnetic poles. In a particular phase, Figure 1a shows the magnetic–flux paths when the sequential magnetic poles are the same, e.g., N–N–N–N and S–S–S–S. The sequential magnetic poles are such that two poles have the same polarity, e.g., S–N–N–S and N–S–S–N in Figure 1b. The opposite of sequential magnetic poles, e.g., N–S–N–S and S–N–S–N, is shown in Figure 1c.

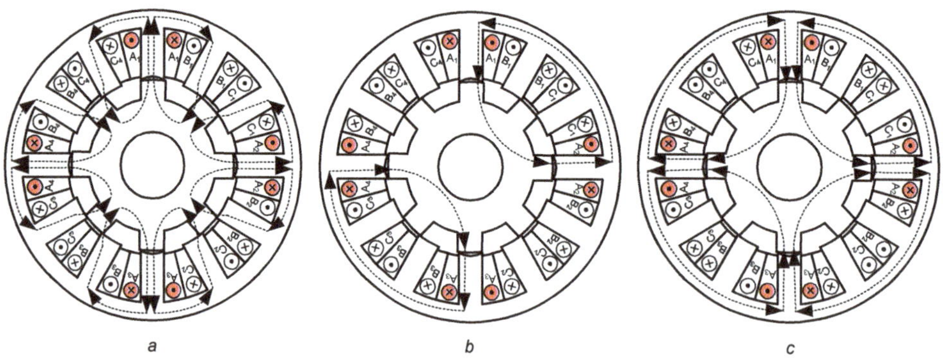

Figure 1. Winding arrangement for single-phase excitation; (**a**) Short-pitched winding 1; (**b**) Short-pitched winding 2; and (**c**) Short-pitched winding 3.

The torque of SRM with short-pitched winding is produced due to the self-inductance variation. The mutual-inductance between the phase windings is ineffective and therefore neglected, resulting in

$$T = \frac{1}{2} i_a^2 \frac{dL_a}{d\theta} + \frac{1}{2} i_b^2 \frac{dL_b}{d\theta} + \frac{1}{2} i_c^2 \frac{dL_c}{d\theta} \tag{1}$$

where the subscripts a, b, and c denote the phase; i, L, and θ are the phase current, self-inductance, and rotor position, respectively.

2.2. Fully-Pitched Winding

Three diagrams of the winding configuration and flux paths in a three-phase 12/8 SRM when two phases (phase A and B) are excited with unipolar operation are shown in Figure 2. In the fully-pitched winding arrangement of the three-phase 12/8 SRM, each coil spans the surrounding three adjacent stator poles. Figure 2a shows the fully-pitched winding diagram when the number of turns per pole is the same as the number of turns per pole of the short-pitched winding, while Figure 2b shows the fully-pitched winding when it is wound with double the number of turns per pole of the short-pitched winding. The fully-pitched winding can operate with either unipolar (double-phase on at a time) or bipolar operation (double-phase on at a time and all three phases on at a time) [5–7].

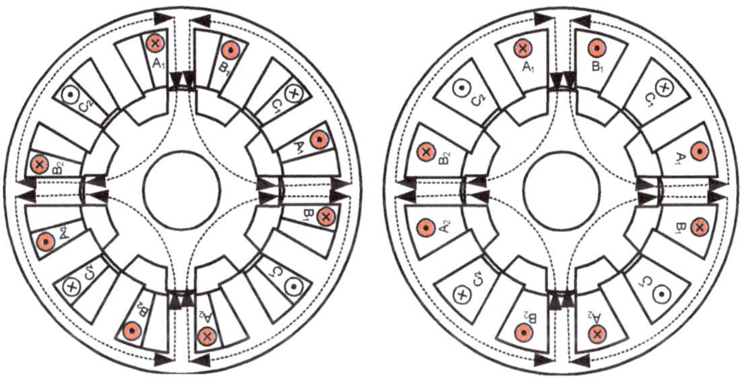

Figure 2. Winding arrangement for double phase excitation; (a) fully-pitched winding 1 and (b) fully-pitched winding 2.

The torque of the fully-pitched winding is only produced due to the rate of change of the mutual-inductance (M) among phases, and torque production can be achieved by exciting two phases simultaneously instead of only one phase in the short-pitched winding. The self-inductance of an individual phase is constant and has the same value as the maximum phase inductance at any rotor position. Therefore, the self-inductance of the fully-pitched winding is independent of the rotor position and is ineffective for torque production. The torque equation of the fully-pitched winding can be expressed by

$$T = i_a i_b \frac{dM_{ab}}{d\theta} + i_b i_c \frac{dM_{bc}}{d\theta} + i_c i_a \frac{dM_{ca}}{d\theta} \tag{2}$$

3. Characteristics of the Winding Arrangement

3.1. Winding Weight and Comparison Based on Equal Conduction Loss

Table 1 shows the dimensions of the three-phase 12/8 SRM. Table 2 shows the winding details of each winding configuration. The number of turns per pole of the short-pitched winding and fully-pitched winding 1 is the same, but it doubles in the fully-pitched winding 2. The number of windings of the short-pitched winding for the three-phase 12/8 SRM generally is the same as the number of stator poles, at 12 coils, as it is a concentrated winding, while the number of windings in fully pitched is decreased to six coils as it is a distributed winding. In terms of the slot-fill factor, which is the ratio of the actual cross–section of copper in a stator slot to the total stator slot area, the fully-pitched winding 1 has the lowest slot–fill factor, while other winding arrangements have the same slot-fill factor.

Table 1. Details and dimensions of the 12/8 SRM.

Parameter	Value
Stator outer diameter [mm]	190
Stator inner diameter [mm]	168
Stator pole arc [mechanical degree]	15
Rotor pole arc [mechanical degree]	16
Shaft diameter [mm]	35
Air gap [mm]	0.3
Stack length [mm]	120
Wire diameter [mm]	0.914 (SWG 20)
Number of strands	2
Rated speed [r/min]	500
Rated torque [Nm]	19
DC link voltage [V]	310
Converter topology	Asymmetric bridge

Table 2. Winding arrangement details.

Winding Arrangement	Number of Turns per Pole	Number of Coils	Excitation Type
Short-pitched 1	97	12	Single phase
Short-pitched 2	97	12	Single phase
Short-pitched 3	97	12	Single phase
Fully-pitched 1	97	6	Double-phase
Fully-pitched 2	194	6	Double-phase

The total winding weight of the fully-pitched winding 2 is the highest at 9.0 kg, while the total weight of windings in the fully-pitched winding 1 is the lowest at 4.5 kg. The slot-fill factor, total weight of windings, and phase resistance of each winding arrangement are shown in Table 3. The total winding weight of all short-pitched winding arrangements is the same at 5.1 kg. The fully-pitched winding 2 has the highest phase resistance value.

Table 3. Electrical and mechanical details.

Winding Arrangement	Slot-Fill Factor	Total Weight of Windings [kg]	Phase Resistance [Ω]
Short-pitched 1	0.37	5.1	1.9
Short-pitched 2	0.37	5.1	1.9
Short-pitched 3	0.37	5.1	1.9
Fully-pitched 1	0.18	4.5	1.7
Fully-pitched 2	0.37	9.0	3.4

In the static machine test, it was thought fair to compare the various designs on the assumption of equal overall winding loss in each situation. Obviously, iron loss will be a key issue in the rotating machine: the influence of magnetic flux reversals on iron loss is explored in a subsequent section, but for a machine rotating at low speed, a comparison based on equal winding loss is sufficient. The winding loss per phase equation in the short-pitched winding SRM is given in

$$\text{Loss}_{\text{SP}} = \frac{1}{3} I_{\text{SP}}^2 R_{\text{SP}} \tag{3}$$

It is assumed that each phase conducts approximately one-third of an electrical cycle. Similarly, in the fully-pitched winding, it can be assumed that each phase conducts approximately two-thirds of one electrical cycle (two-phase excitation), and then the winding loss per phase equation in the fully-pitched winding SRM is given in (4):

$$\text{Loss}_{\text{FULL}} = \frac{2}{3} I^2_{\text{FULL}} R_{\text{FULL}} \tag{4}$$

where the subscripts SP and FULL denote the short-pitched winding and fully-pitched winding, respectively; I and R are the conduction current and phase resistance, respectively.

To compare the five winding arrangements of a three-phase 12/8 SRM based on the same conduction loss for the same motor structure, the phase current of each winding arrangement is selected according to Equations (3) and (4) for the short-pitched and fully-pitched winding, respectively. Table 4 shows the phase current of each winding configuration based on equal conduction loss. In the Finite Element Analysis (FEA) model, the phase current of each winding arrangement in Table 4 is employed to simulate phase torque production.

Table 4. Phase winding current comparison for the same conduction loss.

Winding Arrangement	Phase Winding Current at the Same Conduction Loss [A]
Short-pitched 1, 2, 3	15
Fully-pitched 1	11.27
Fully-pitched 2	7.98

The excitation methods under unipolar operation of the winding arrangement are shown in Table 2. The short-pitched winding arrangements are excited with single-phase excitation at a time, while the fully-pitched winding arrangements are excited with double-phase excitation at a time.

3.2. Core Loss Based on Flux Reversals

The core losses consist of three parts: hysteresis loss, eddy current loss, and excess loss as shown in Equation (5). The calculation of the core loss per unit volume (P_v) in the frequency domain is widely used for ferromagnetic material [17] as it can be expressed as Equation (6). k_{hys}, k_{eddy}, k_{excess} are the coefficients and β is the parameter of the core loss which are obtained from manufacturer data or experimental testing based on the curve–fitting method. B_m is the peak magnetic flux density and f is the frequency. This equation is suitable for sinusoidal flux density waveforms, and it does not consider the minor loop in the main hysteresis loop.The frequency of the flux is determined by the phase switching frequency:

$$P_v = P_{hys} + P_{eddy} + P_{excess} \tag{5}$$

$$P_v = k_{hys} f B_m^{\beta} + k_{eddy}(f B_m)^2 + k_{excess}(f B_m)^{1.5} \tag{6}$$

For SRM, the flux, which is distributed on the machine, is nonsinusoidal waveforms. Using the conventional equation may lead to large errors in the core loss calculation. Then, the core loss can be analyzed in the time domain [18], which is suitable for the nonsinusoidal flux waveform, and it can be given by Equations (7)–(9):

$$P_{hys}(t) = k_{hys} H_{irr} \frac{dB}{dt} \tag{7}$$

$$P_{eddy}(t) = k_{eddy} \frac{1}{2\pi^2} \left[\frac{dB}{dt}\right]^2 \tag{8}$$

$$P_{excess}(t) = k_{excess}\frac{1}{C_e}\left[\left|\frac{dB}{dt}\right|\right]^{1.5} \tag{9}$$

H_{irr} is evaluated based on an equivalent elliptical loop and C_e is the numerical integration parameter as described in [18]. From the core loss equation in the time domain, it can be seen that the main contributory impact of the core loss is the rate of change of the flux density with time; dB/dt. Modifying winding arrangement can change the direction of the flux path, which affects the flux density of the machine. The phenomenon of flux reversals in the stator core back is one of the causes that affects the variation of the flux density in the SRM [14–16]. Therefore, minimizing the number of flux reversals occurring on the stator core back can decrease the core loss in SRM.

4. Selection of Winding Arrangement

4.1. Average Static Phase Torque

The effective static phase torque is generated according to the inductance profile of each winding arrangement, as shown in Figure 3. The effective torque regions, which are the positive torque production zones of the three-phase 12/8 SRM, are 22.5 mechanical degrees. Figure 4 shows the effective static phase torque profiles of each winding arrangement based on equal conduction loss. From Figure 4, the fully-pitched winding 2, which is double-phase excitation, can produce the highest peak torque, at 44.08 Nm. For short-pitched winding arrangements, the short-pitched winding 1, which requires single phase excitation, can generate a peak torque of 35.18 Nm, but it produces negative torque at some position, and the short-pitched windings 2 and 3, which require single phase excitation, can generate a peak torque of 40.47 Nm and 41.4 Nm, respectively. In other fully-pitched winding arrangements, the fully-pitched winding 1, which uses double-phase excitation, can produce a peak torque of 29.3 Nm.

It can be observed in the short-pitched winding 1, which produces some negative torque due to its inductance profile having an early decrease slope before going to an aligned position at 22.5 degrees. The inductance profile of the fully-pitched winding 2 has a high slop caused by the high number of winding turns at 194 compared with other winding arrangements. Table 5 lists the average value of the effective torque on the basis of the same conduction loss according to the torque profiles in Figure 4. It is clear that the fully-pitched winding 2 can generate the highest average torque, while the short-pitched winding 1 can produce the lowest average torque.

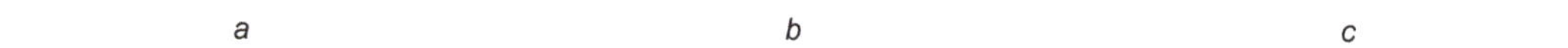

Figure 3. Inductance profiles of a three-phase 12/8 SRM at different winding arrangements; (**a**) Short-pitched winding; (**b**) Short-pitched winding 2; and (**c**) Fully-pitched winding.

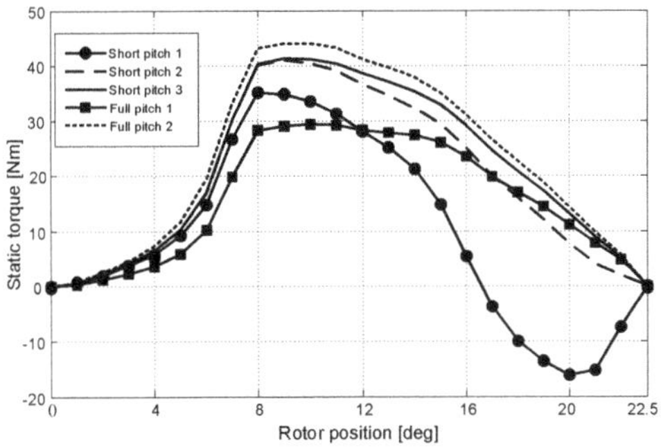

Figure 4. Effective static torque of a three-phase 12/8 SRM at different winding arrangements on the basis of equal conduction loss.

Table 5. Average static torque at the same conduction loss.

Winding Arrangement	Static Torque [Nm]	Torque/Copper Weight [Nm/kg]
Short-pitched 1	9.8	1.9
Short-pitched 2	19.6	3.8
Short-pitched 3	21.6	4.2
Fully-pitched 1	15.9	3.5
Fully-pitched 2	23.4	2.6

4.2. Number of Flux Reversals on Stator Yoke

In this paper, the number of flux reversals is considered only on the stator yoke. The leakage flux is neglected. The yoke of the 12/8 SRM stator is divided into 12 segments. Each segment of the stator yoke experiences flux passing through it when the phase winding is excited. With different phases of winding excitation, the direction of the flux path in each stator yoke segment is varied.

As mentioned in the above discussion, the rate of change of the flux density with time significantly affects the core loss of the SRM stator. The number of flux reversals is one of the causes that involves the rate of change of the flux density on the stator core. In this paper, the number of flux reversals occurring on the stator yoke is observed by counting the number of changes in flux path direction in each stator segment when the phase windings are sequentially excited in one revolution. The number of flux reversals in the stator yoke can be given as

$$N_{FR} = \left(\sum_{i=1}^{n} N_{FD}(i) \right) * N_s \tag{10}$$

where N_{FR} is the total number of flux reversals on the yoke of the stator, N_{FD} is the number of changing flux path directions in each segment of the stator yoke when the phase windings are sequentially excited (A–B–C–A) , N_s is the number of stator poles, and n is the number of stator yoke segments.

Table 6 shows the number of flux reversals on the stator yoke in the different winding configurations. The short-pitched winding 3 and the fully-pitched winding have the highest number of flux reversals on the stator yoke at 288, while the short-pitched winding 2 does not have the number of flux reversals on the stator yoke. The short-pitched winding 1 still has a high number of flux reversals on the yoke of the stator at 240.

Table 6. Number of flux reversals in the stator yoke in 1 revolution.

Winding Arrangement	Number of Flux Reversals
Short-pitched 1	240
Short-pitched 2	0
Short-pitched 3	288
Fully-pitched 1	288
Fully-pitched 2	288

4.3. *Consideration of Winding Selection*

In terms of application, the ventilation exhaust fan for poultry farming requires a motor to run at a low rotational speed, have a high starting torque output, and be robust enough for the hard environment in a farming system. In terms of motor manufacturing, simple structure, no complexity in the assembly process, and low copper consumption for motor winding are the key factors for the candidate motor. From both points of view, SRM is a suitable choice for this application. Rated speed and load torque, including voltage range and converter topology of the SRM in this research, are shown in Table 1.

The winding selection of SRM is determined by the characteristics of creating the high torque capacity to achieve a high–efficiency motor and using low copper, resulting in cost savings in the manufacture. Tables 2 and 3 are used for considering the complexity of the winding assembly process and copper consumption in each winding configuration. The fully-pitched winding has a small number of coils at 6 coils, while the short-pitched winding has 12 coils. However, in terms of the manufacturing process, the short-pitched winding with a concentrated winding gives more advantages, such as the winding of a coil around a single–pole and short end–turn of coil compared with the fully-pitched (winding of a coil around three stator poles), which is similar to a distributed winding. Furthermore, the fully-pitched winding 2 uses more copper at 9 kg compared with any other type of winding arrangement.

To compare the torque performance, the ratio of the static phase torque per copper weight is applied for consideration as the stator and rotor core are the same weight. The static phase torque results based on equal conduction loss were obtained from the FEA model of SRM, which applied the phase current value in Table 4 calculating based on equal conduction loss according to Equations (3) and (4) for the short-pitched and fully-pitched winding, respectively.

As shown in Table 5, the ratio of the static phase torque per copper weight of the short-pitched windings 2 and 3 is high compared with other types of winding configuration. In terms of motor efficiency, the number of flux reversals of the short-pitched winding 2 is the lowest. On the other hand, the short-pitched winding 3 has a high number of flux reversals, as shown in Table 6. Therefore, two candidates for the winding arrangement of the 12/8 SRM are the short-pitched windings 2 and 3. In the next analysis, the dynamic simulation analysis of each winding pattern will be discussed.

5. Dynamic Simulation

The dynamic simulation in this section is performed by JMAG finite element analysis software. The control parameters of the 12/8 SRM in the dynamic simulation are the same. The advance turn–on angle is 0 degrees, the conduction period is 22.5 degrees in the short-pitched winding, the conduction period is 30 in the fully-pitched winding, and the rotational speed is 500 r/min. A conventional hysteresis current control is applied and the maximum current is 9 A. The five winding configurations of the 12/8 SRM are simulated. Average torque and torque ripple are the essential considerations in this simulation.

Figure 5 shows the flux density and flux path of the dynamic simulation. The result of the dynamic simulation is shown in Figures 6 and 7 and comparative results in Table 7. The dynamic torque performance of the double-phase excitation of the fully-pitched winding is shown in Figure 7. The torque ripple rate of the fully-pitched winding 1 and the fully-pitched winding 2 is 1.3 and 1.0, respectively. The average torque of the fully-pitched winding is the highest, at 36.55 Nm. However, with double-phase excitation and high phase resistance of the fully-pitched winding, it produces more copper losses compared with the other winding arrangements.

Table 7. Comparison of dynamic simulation performance.

Winding Arrangement	Average Torque [Nm]	Torque Ripple Rate
Short-pitched 1	3.54	10.28
Short-pitched 2	11.81	2.25
Short-pitched 3	19.0	0.66
Fully-pitched 1	17.12	1.30
Fully-pitched 2	36.55	1.00

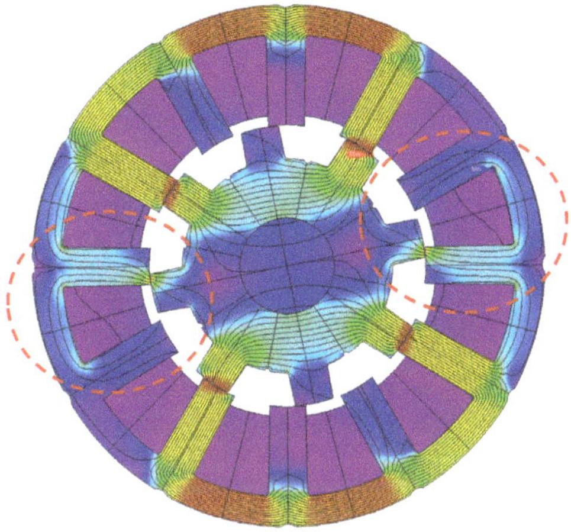

Figure 5. Leakage-flux path of the short-pitched winding 2 when operates at high current densities.

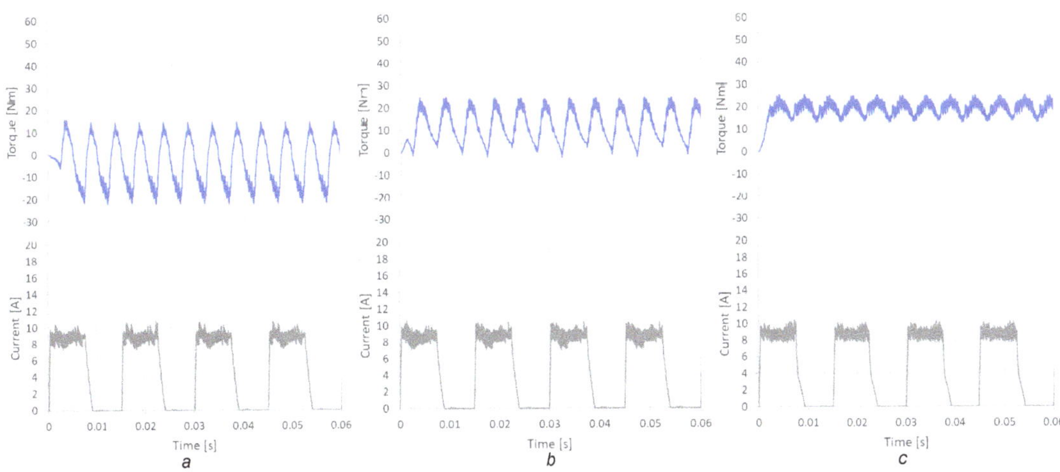

Figure 6. Torque and current waveform (Phase A) for single-phase excitation; (**a**) Short-pitched winding 1; (**b**) Short-pitched winding 2; and (**c**) Short-pitched winding 3.

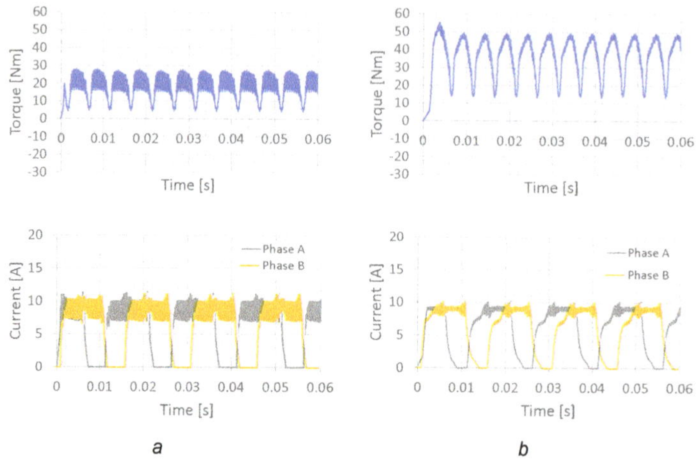

Figure 7. Torque and current waveform for double-phase excitation; (**a**) Fully-pitched winding 1 and (**b**) Fully-pitched winding 2.

The dynamic torque performance of the single-phase excitation in Figure 6 shows that the short-pitched winding 1 has a dramatically torque ripple rate of 10.28, and the average torque is lowest at 3.54 Nm. The short-pitched winding 2 has a torque ripple rate of 2.25 and the average torque decreases to 11.81 Nm. compared with the short-pitched winding 3 having a low torque ripple rate at 0.66 and average torque at 19 Nm. To describe this phenomenon, Figure 5 shows the flux density and flux path of the short-pitched winding 2. The simulation results of the short-pitched winding 2 show that the leakage–flux paths appear on the adjacent stator poles when operating at high torque density. Therefore, the number of flux reversals in the stator yoke increases and generates more negative torque during phase commutation. The leakage–flux paths are shown in the red dot circle in Figure 5. The short-pitched winding 2 has good torque performance and reduces the number of flux reversals when operating at low torque density, as was revealed in [15]. For this reason, the average torque of the short-pitched winding 2 decreases, and the performance of the SRM drops to 76.63%, while the short-pitched winding 3 has a higher average torque and efficiency at 84.9%. However, the core loss of the short-pitched winding 2 is still lower than the short-pitched winding 3. Table 8 reveals comparative results of dynamic simulation performance between short-pitched 2 and short-pitched 3 in terms of core loss at the stator and efficiency performance. For ventilation fan applications, the torque ripple has a significant impact on this application. A high torque ripple will lead to increased vibration and acoustic noise.

Table 8. Comparison of motor performance between short-pitched 2 and short-pitched 3.

	Short-Pitched 2	**Short-Pitched 3**
Average torque [Nm]	11.81	19.0
Torque ripple rate	2.25	0.66
Core loss at stator [W]	37.67	38.44
Efficiency [%]	76.63	84.9

From the dynamic simulation results, it can be concluded that the short-pitched winding 3 is suitable for 12/8 SRM in terms of cost-saving, simple manufacturing process, torque ripple, and motor performance.

6. Flux Reversal Reduction

Although the short-pitched winding 3 gives a better performance for 12/8 SRM, the number of flux reversals of this winding configuration is still high. This paper proposes an analysis method that can reduce the flux reversals on the stator yoke. The short-pitched winding 3 is divided into Design A and Design B as follows:

- In Design A, as shown in Figure 8, the sequential magnetic poles in a clockwise direction are SNSN when excited phase A, NSNS when excited phase B, and NSNS when excited phase C.
- In Design B, as shown in Figure 9, the sequential magnetic poles in a clockwise direction are SNSN when excited phase A, NSNS when excited phase B, and SNSN when excited phase C.

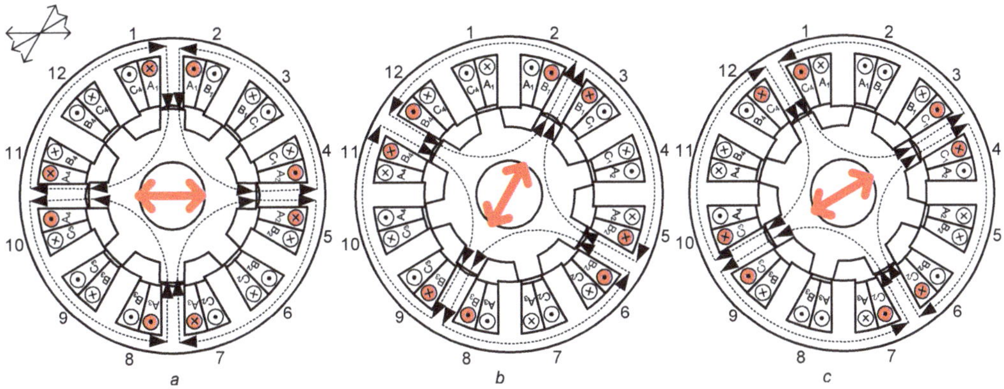

Figure 8. Excitation pattern and flux-path of the short-pitched winding 3 in Design A; (**a**) Phase A is excited; (**b**) Phase B is excited; and (**c**) Phase C is excited.

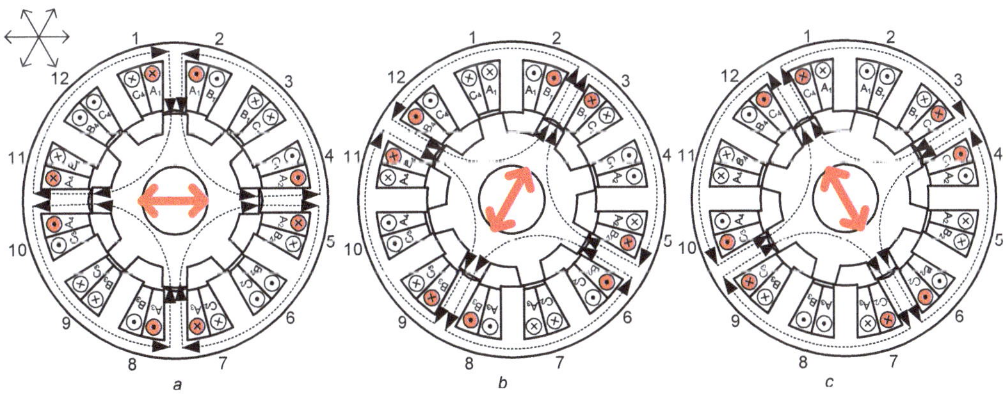

Figure 9. Excitation pattern and flux-path of the short-pitched winding 3 in Design B; (**a**) Phase A is excited; (**b**) Phase B is excited; and (**c**) Phase C is excited.

The number of flux reversals in each segment of the stator yoke in the clockwise (A–B–C–A) and counter-clockwise (A–C–B–A) direction of Design A is illustrated in Table 9. To calculate the total number of flux reversals in one revolution, Equation (10) is used. The total number of flux reversals on the yoke of the stator is 192 in both directions (clockwise and counter–clockwise). Using the same procedure, Design B has a total number of flux reversals in one revolution of 288 in both directions (clockwise and counter–clockwise) as shown in Table 10. Table 11 reveals the results of the comparison of the short-pitched winding 3 with

Designs A and B. The results show that the stator core loss is decreased in the short-pitched winding 3 with Design A.

Table 9. Number of flux reversals in Design A.

Stator Yoke Sector	Flux-Linkage Direction (Clockwise)	No. of Reverse Direction (Clockwise)	Flux-Linkage Direction (Counter Clockwise)	No. of Reverse Direction (Counter Clockwise)
1	→←←→	2	→←←→	2
2	←←←←	0	←←←←	0
3	←←→←	2	←→←←	2
4	←→→←	2	←→→←	2
5	→→→→	0	→→→→	0
6	→→←→	2	→←→→	2
7	→←←→	2	→←←→	2
8	←←←←	0	←←←←	0
9	←←→←	2	←→←←	2
10	←→→←	2	←→→←	2
11	→→→→	0	→→→→	0
12	→→←→	2	→←→→	2
	Total	16	Total	16

Table 10. Number of flux reversals in Design B.

Stator Yoke Sector	Flux-Linkage Direction (Clockwise)	No. of Reverse Direction (Clockwise)	Flux-Linkage Direction (Counter Clockwise)	No. of Reverse Direction (Counter Clockwise)
1	→→←→	2	→←→→	2
2	←→←←	2	←←→←	2
3	←→→←	2	←→→←	2
4	←←→←	2	←→←←	2
5	→←→→	2	→→←→	2
6	→←←→	2	→←←→	2
7	→→←→	2	→←→→	2
8	←→←←	2	←←→←	2
9	←→→←	2	←→→←	2
10	←←→←	2	←→←←	2
11	→←→→	2	→→←→	2
12	→←←→	2	→←←→	2
	Total	24	Total	24

Table 11. Comparison of winding performance short-pitched winding 3 Design A (SP 3 Design A) and short-pitched winding 3 Design B (SP 3 Design B).

	SP 3 Design A	SP 3 Design B
Average torque [Nm]	18.9	19.0
Torque ripple rate	0.66	0.66
Core loss at stator [W]	36.36	38.44
Efficiency [%]	85.14	84.9

Figure 10 shows the comparative results of core loss between short-pitched 3 Design A and Design B at each rotational speed and current level, respectively. It can be seen that Design A can reduce the core loss by around 7–9% at the rotational speed of 300–400 r/min with low current and around 10–13% at the rotational speed of 500 r/min with high current.

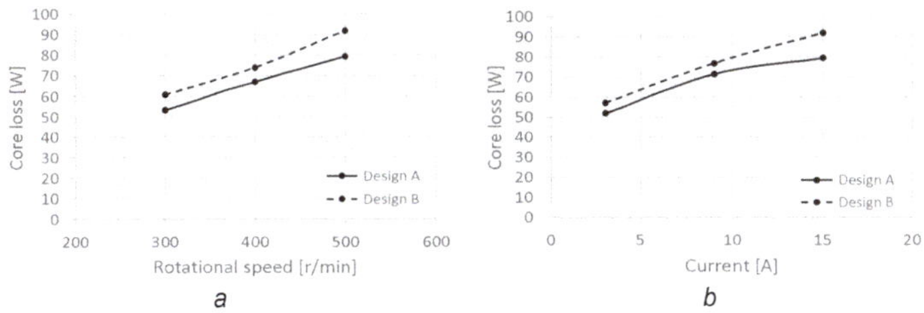

Figure 10. Core loss of the short-pitched winding 3 Design A and Design B; (**a**) at different rotational speeds and (**b**) at different current levels.

7. Experimental Results

The three-phase 12/8 SRM prototype was built for ventilation fan applications using the existing induction motor frame size of 112 M, including cover, motor frame, and shaft according to the IEC standard. Therefore, the maximum outer diameter is 190 mm and the shaft diameter is 35 mm. Figure 11 reveals the example of a single coil winding on the stator pole. The winding arrangement of the SRM prototype is the short-pitched winding 3 Design A, and the winding diagram is shown in Figure 12. Figure 13 illustrates the completed prototype of the 12/8 SRM.

Figure 11. One coil winding on the stator pole.

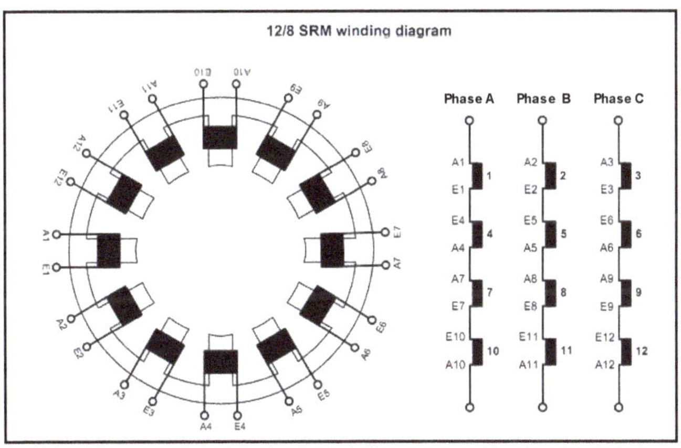

Figure 12. Winding diagram of the 12/8 SRM prototype.

Figure 13. Stator, rotor, and completed 12/8 SRM prototype.

The experimental system setup is shown in Figure 14. The rotor of the three-phase 12/8 SRM was connected to the rotor of the load machine (20 kW Siemens induction motor) via a flexible coupling and torque transducer. The turn-on and turn-off angles of SRM are 0 and 22.5 deg. The DC link voltage was 310 V. The experimental test was implemented at 500 r/min. The rated load torque output of 19.4 Nm of the load machine was applied to the 12/8 SRM. The current and voltage waveform of the experimental test are shown in Figure 15a,b at the rotational speed of 600 r/min with load torque at 18.5 Nm of the load machine. It can be seen that the SRM prototype running at 600 r/min can carry less load torque due to the influence of voltage back electromotive force, which increases with rotational speed. Therefore, phase current cannot reach the level of the desired current, leading to low torque capability. However, when the phase current decreases, the loss of the copper also reduces. For this reason, the motor performance of 600 r/min gives high efficiency.

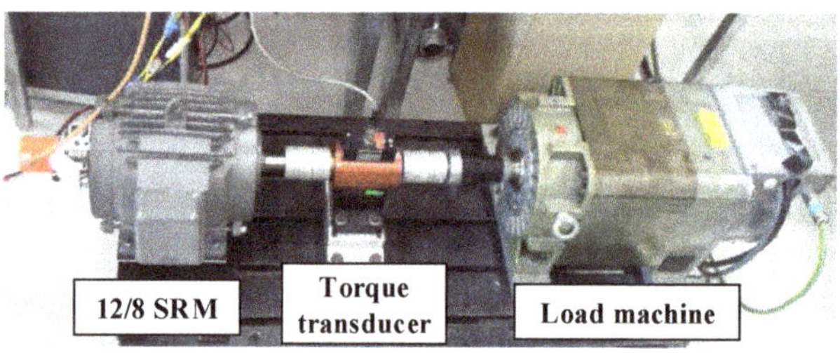

Figure 14. Experimental test rig.

Figure 16 shows the experimental results of torque vs. speed and efficiency vs. speed curves with different rotational speeds between 300–1500 r/min. The maximum torque of the SRM prototype is 21.8 Nm at 300 r/min. At the rated speed of 500 r/min, the SRM prototype can carry torque at 19.4 of load torque which covers the desired rated torque, and the efficiency of the motor is 82.5%. The maximum efficiency of the SRM prototype can reach 83.1% at 600 r/min with 18.5 Nm of load torque.

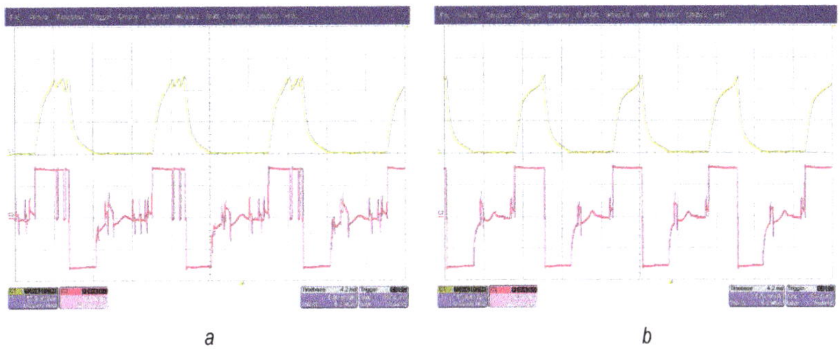

Figure 15. Phase A current (**Top**) and voltage (**Bottom**) waveform of the 12/8 SRM prototype using SP 3 Design A winding configuration; (**a**) at 500 r/min with 19.4 Nm of load torque and (**b**) at 600 r/min with 18.5 Nm of load torque.

The comparative results at rated conditions between the simulation model and experimental test are illustrated in Table 12. The SRM prototype gives a torque output of 19.4 Nm close to the simulation model, which has 18.9 Nm. However, the efficiency of the SRM prototype decreased to 82.5%, while the simulation model could achieve 85.14%. In the experimental test, the maximum current of the hysteresis current control is higher than that of the simulation, so this will lead to an increase in the copper loss of the system. To achieve the rated torque of 19.0 Nm, the SRM prototype requires more phase current than that in the simulation model due to the imperfections during the manufacture and assembly process, such as errors in air gap length, non-uniform air gap, and eccentricity error, etc.

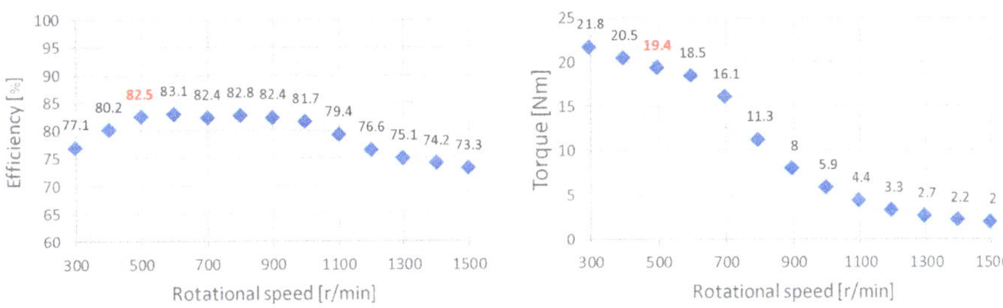

Figure 16. Efficiency and torque at different rotational speeds.

Table 12. Experimental results.

	SP 3 Design A FEA	SP 3 Design A Exp
Average torque [Nm]	18.9	19.4
Efficiency [%]	85.14	82.5
Aligned inductance [mH]	218.57	215.54
Unaligned inductance [mH]	21.86	27.37

8. Conclusions

This research has introduced the criteria to consider for choosing the winding configuration of three-phase 12/8 SRM as follows:

In the analysis of the static characteristic: (1) the ratio value of static phase torque (based on equal conduction loss) per weight of copper winding needs to be considered for selecting winding configuration to produce high efficiency and cost-saving in the production process; (2) the number of flux reversals on the stator yoke of the SRM, winding configuration having a low number of flux reversals can reduce the core loss.

In the analysis of the dynamic characteristics, the torque ripple is the main parameter that each winding configuration needs to consider, especially since the torque ripple has a significant impact on the ventilation fan application.

The comparison of the winding arrangements for the same motor structure reveals that the total winding weight of the fully-pitched winding 2 is the highest because of the increased end windings. The comparison of the average torque results on the basis of equal conduction loss shows that the fully-pitched winding 2 with double-phase unipolar excitation and the short-pitched winding 3 with single phase unipolar excitation are effective for enhancing the torque density. The current level of the fully-pitched winding 2 has to be reduced to attain the same conduction loss where average output torque is still higher than any other winding arrangement.

The consideration of winding selection is discussed based on cost-saving, the complexity of the manufacturing process, and machine performance, including torque output and torque ripple. The short-pitched winding 3 gives significantly better performance in terms of the ratio of the static phase torque per copper weight compared with other winding configurations. For initial consideration of core loss in SRM, the counting of the number of flux reversals on the stator yoke is proposed. The simulation results confirm that minimizing the number of flux reversals in the stator yoke by modifying the winding polarities can decrease the core loss of the SRM. Finally, the 12/8 SRM prototype was built and tested (load test and characteristic test).

Author Contributions: Conceptualization, R.P. and K.T.; methodology, R.P. and K.T.; validation, R.P., S.K., P.J., P.S. and K.T.; formal analysis, R.P. and K.T.; investigation, R.P., P.S. and K.T.; resources, S.K., P.J. and P.S.; writing—original draft preparation, R.P.; writing—review and editing, R.P., S.K., P.J., P.S. and K.T.; visualization, R.P. and S.K.; supervision, K.T. All authors have read and agreed to the published version of the manuscript.

Funding: This research received no external funding.

Institutional Review Board Statement: Not applicable.

Informed Consent Statement: Not applicable.

Data Availability Statement: Not applicable.

Conflicts of Interest: The authors declare no conflict of interest.

Nomeclatures

N	north magnetic pole
S	south magnetic pole
T	torque
i	phase current
θ	rotor position
L	self-inductance
M	mutual-inductance
SP	short-pitched winding
FULL	fully-pitched winding
I	conduction current
R	phase resistance
P_v	core losses
P_{hys}	hysteresis loss
P_{eddy}	eddy current loss
P_{excess}	excess loss
k_{hys}	hysteresis loss coefficient
k_{eddy}	eddy current loss coefficient
k_{excess}	excess loss coefficient

β	core loss parameter based on the curve-fitting method
B_m	peak magnetic flux density
B	magnetic flux density
H_{irr}	evaluated value based on an equivalent elliptical loop [18]
C_e	numerical integration parameter [18]
NFR	total number of flux reversals on stator yoke
NFD	number of changing flux path direction in each segment of stator yoke
N_s	number of stator poles
n	number of stator yoke segments

References

1. Miller, T.J.E. *Switched Reluctance Motors and Their Control*; Magna Physics Publishing and Clarendon Press: Oxford, UK, 1993.
2. Krishnan, R. *Switched Reluctance Motor Drives: Modeling, Simulation, Analysis, Design, and Applications*; CRC Press: Boca Raton, FL, USA, 2001.
3. Widmer, J.D.; Martin, R.; Mecrow, B.C. Optimization of an 80-kW Segmental Rotor Switched Reluctance Machine for Automotive Traction. *IEEE Trans. Ind. Appl.* **2015**, *51*, 2990–2999. [CrossRef]
4. Rahman, M.S.; Lukman, G.F.; Hieu, P.T.; Jeong, K.I.; Ahn, J.W. Optimization and Characteristics Analysis of High Torque Density 12/8 Switched Reluctance Motor Using Metaheuristic Gray Wolf Optimization Algorithm. *Energies* **2021**, *14*, 2013. [CrossRef]
5. Mecrow, B. Fully pitched-winding switched-reluctance and stepping-motor arrangements. *IEE Proc. B (Electr. Power Appl.)* **1993**, *140*, 61–70. [CrossRef]
6. Barrass, P.G.; Mecrow, B.C.; Clothier, A.C. Bipolar operation of fully-pitched winding switched reluctance drives. In Proceedings of the 1995 Seventh International Conference on Electrical Machines and Drives (Conf. Publ. No. 412), Durham, UK, 11–13 September 1995; pp. 252–256. [CrossRef]
7. Mecrow, B.C. New winding configurations for doubly salient reluctance machines. *IEEE Trans. Ind. Appl.* **1996**, *32*, 1348–1356. [CrossRef]
8. Li, Y.; Tang, Y. Switched reluctance motor drives with fractionally-pitched winding design. In Proceedings of the PESC97, Record 28th Annual IEEE Power Electronics Specialists Conference, Formerly Power Conditioning Specialists Conference 1970–71. Power Processing and Electronic Specialists Conference 1972, St. Louis, MO, USA, 27 June 1997; Volume 2, pp. 875–880. [CrossRef]
9. Tang, Y. Switched reluctance motor with fractionally pitched windings and bipolar currents. In Proceedings of the Conference Record of 1998 IEEE Industry Applications Conference, Thirty-Third IAS Annual Meeting (Cat. No. 98CH36242), St. Louis, MO, USA, 12–15 October 1998; Volume 1, pp. 351–358. [CrossRef]
10. Chen, H.; Song, Q. Windings arrangement of a three-phase switched reluctance machine. In Proceedings of the IEEE International Electric Machines and Drives Conference, IEMDC'03, Madison, WI, USA, 1–4 June 2003; Volume 3, pp. 1665–1668. [CrossRef]
11. Pupadubsin, R.; Chayopitak, N.; Karukanan, S.; Champa, P.; Somsiri, P.; Tungpimolrut, K. Comparison of winding arrangements of three phase switched reluctance motor under unipolar operation. In Proceedings of the 2012 15th International Conference on Electrical Machines and Systems (ICEMS), Sapporo, Japan, 21–24 October 2012; pp. 1–4.
12. Augustine, M.; Balaji, M.; Kamaraj, V. Characteristics Assessment of Switched Reluctance Motor with Segmented Rotor. In Proceedings of the 2019 IEEE 1st International Conference on Energy, Systems and Information Processing (ICESIP), Chennai, India, 4–6 July 2019; pp. 1–6. [CrossRef]
13. Prabhu, S.; Balaji, M.; Kamaraj, V. Analysis of two phase switched reluctance motor with flux reversal free stator. In Proceedings of the 2015 IEEE 11th International Conference on Power Electronics and Drive Systems, Sydney, NSW, Australia, 9–12 June 2015; pp. 320–325. [CrossRef]
14. Oh, S.; Krishnan, R. Two-Phase SRM With Flux-Reversal-Free Stator: Concept, Analysis, Design, and Experimental Verification. *IEEE Trans. Ind. Appl.* **2007**, *43*, 1247–1257. [CrossRef]
15. Lobo, N.S.; Swint, E.; Krishnan, R. M-Phase N-Segment Flux-Reversal-Free Stator Switched Reluctance Machines. In Proceedings of the 2008 IEEE Industry Applications Society Annual Meeting, Edmonton, AB, Canada, 5–9 October 2008; pp. 1–8. [CrossRef]
16. Seshadri, A.; Lenin, N.C. Review based on losses, torque ripple, vibration and noise in switched reluctance motor. *IET Electr. Power Appl.* **2020**, *14*, 1311–1326. [CrossRef]
17. Bertotti, G. General properties of power losses in soft ferromagnetic materials. *IEEE Trans. Magn.* **1988**, *24*, 621–630. [CrossRef]
18. Lin, D.; Zhou, P.; Fu, W.N.; Badics, Z.; Cendes, Z.J. A dynamic core loss model for soft ferromagnetic and power ferrite materials in transient finite element analysis. *IEEE Trans. Magn.* **2004**, *40*, 1318–1321. [CrossRef]

energies
MDPI

Article

Performance Improvement of a Switched Reluctance Motor and Drive System Designed for an Electric Motorcycle

Seubsuang Kachapornkul, Ruchao Pupadubsin, Pakasit Somsiri, Prapon Jitkreeyarn and Kanokvate Tungpimolrut *

Machines and Power Conversions Research Team, National Electronics and Computer Technology Center, 112 Thailand Science Park, Phahonyothin Rd., Khlong Nueng, Khlong Luang 12120, Pathumthani, Thailand; seubsuang.kachapornkul@nectec.or.th (S.K.); ruchao.pupadubsin@nectec.or.th (R.P.); pakasit.somsiri@nectec.or.th (P.S.); prapon.jitkreeyarn@nectec.or.th (P.J.)
* Correspondence: kanokvate.tungpimolrut@nectec.or.th

Abstract: In this paper, the implementation of a switched reluctance motor (SRM) and drive system for the propulsion system of a two-seat electric motorcycle is described. The overall design focuses on the required vehicle speed, acceleration, driving distance, and overall system cost, as well as reliability. The performance of the three-phase 6/4 pole (six-stator pole and four-rotor pole) and four-phase 8/6 pole (eight-stator pole and six-rotor pole) are investigated and compared by static performance analysis and dynamic performance analysis. Their performance is further investigated by finite element analysis. The indirect torque controller in a drive system for optimal torque and efficiency operation is also mentioned. A methodology for rotor position detection and its hardware implementation are also proposed. The designed 3.5 kW three-phase 6/4 pole SRM and its drive system were constructed and tested on the test bench. A maximum efficiency of about 82% could be achieved for the SRM and drive system. It was also installed on a 120-cc electric motorcycle, and the vehicle's performance was also validated by on-road and dynamometer testing. The maximum vehicle speed reached was 82 km/h, and a cruising distance of about 98 km at a constant speed of 40 km/h was measured.

Keywords: electric motorcycle; switched reluctance motors; indirect torque control

Citation: Kachapornkul, S.; Pupadubsin, R.; Somsiri, P.; Jitkreeyarn, P.; Tungpimolrut, K. Performance Improvement of a Switched Reluctance Motor and Drive System Designed for an Electric Motorcycle. *Energies* **2022**, *15*, 694. https://doi.org/10.3390/en15030694

Academic Editor: Vladimir Prakht

Received: 21 November 2021
Accepted: 14 January 2022
Published: 18 January 2022

1. Introduction

A motorcycle is the most popular vehicle in Thailand and ASEAN countries. In Thailand, more than 20 million units, or 50% of all on-road vehicles, registered to the Department of Transportation are motorcycles. Furthermore, there has been a rapidly increasing trend in the number of motorcycles during the pandemic for delivery services and urban commuting, which has caused more serious air pollution and health problems. In order to tackle this problem, converting existing old motorcycles or using new electric motorcycles is considered to be an interesting policy of government agencies [1]. Recently, some electric motorcycles and small electric scooters have been exported. The propulsion systems of those vehicles mainly use the motors that contain permanent magnets inside, which causes some difficulties for local mass production and uncontrollable price fluctuation. From the literature [2,3], the switched reluctance motor (SRM) is one type of motor that does not contain any rare earth materials or permanent magnets inside. It presents many advantages, opportunities, and challenges over the other type of electric motors for electric vehicle applications. Various structures and winding configurations of the SRM as well as various driving topologies have been reported [4]. It has been widely studied and installed for traction drives in various hybrid [5–7] and electric vehicles, including bicycles [8], forklifts [9], and light electric vehicles [10]. The design, production, and verification of a complete running prototype based on SRM wheel hub drive train has been reported [11]. A control method in the drive systems of electric motorcycles using SRMs [12] has also been previously reported. Some previous publications focused only on the performance improvement of the SRM drive system by optimizing the turn on and turn off angle at each rotating speed and vehicle load torque [13], or by simultaneously adjusting the geometry and commutation angles [14].

The SRM has a special structure and several advantages, such as rugged and low-cost construction, a simple stator and rotor structure, an easy cooling system, high reliability, and good performance over a wide speed range. These characteristics make it a low-cost and rugged alterative for industry applications, a worthy competitor to other drive systems, and suitable for various applications. Additionally, its performance can be easily adjusted for various load profiles. Thus, it is also one of the most interesting applications for electric and hybrid vehicle drive systems. Unlike permanent magnet machines, the SRM is easy to manufacture with a cheaper cost, since it does not need a permanent magnet. Moreover, the SRM is very suitable for harsh and high-temperature environments. For the sake of safety in a traction drive system [2], very high torque in the low-speed region is required. Furthermore, for the wide constant power region during high vehicle speeds, the power train must be mostly operated in the constant power region. It is reported in [3] that the three-phase 6/4 pole SRM has an overload capacity and constant power region with a longer range, which is suitable for the traction drive application. However, it has a bigger torque ripple and some vibration. Therefore, some papers overlook this problem by using a four-phase motor [7] or increasing the number of stator and rotor poles [5], such as 8/6, 12/8, or 6/10 pole [8] SRMs. However, the increasing number of components for increasing phases also causes the drive circuit to be more expensive and have more complexity.

Until now, there has been a variety of literature on the design of SRMs. The design in each paper focuses on different points of view. In the case of electric motorcycle applications, it is very challenging to design a drive system and a high torque density motor (per volume or weight) with limited space, low cost, and high reliability in order to meet high performance and high efficiency requirements and to make it acceptable for many customers. In this paper, the first task to be accomplished is to design a motor to satisfy the performance specifications and requirements of the vehicle. A high starting torque motor and drive system was designed in order to reduce the acceleration and deceleration time to an acceptable value. The motor design mainly focuses on the starting torque, whereas the other criteria, such as the good performance and high efficiency of the motor for longer driving distances, are also taken into consideration. The geographical dimensions for the design guidelines of the stator, rotor core, and winding parameter for the four-phase 8/6 pole and three-phase 6/4 pole SRM were considered. The comparison of both types of SRMs was investigated by the simulation results using a static performance analysis tool. The dynamic performance analysis of the four-phase 8/6 pole and three-phase 6/4 SRMs, including the control methodology in the drive system, were also investigated using the dynamic simulation tool. Furthermore, the developed torque and flux density of different winding configurations for the three-phase 6/4 SRM were also investigated. In order to implement a low-cost and high-reliability drive system, an indirect toque control method for optimal torque and efficiency operation and higher accuracy of rotor position detection based on a combination of a magnetic sensing circuit and an initial rotor position estimation method are proposed. The construction and the measured performance on the test bench of the selected three-phase 6/4 pole SRM and drive system are described. Finally, the measured results for the vehicle performance on a dynamometer and during on-road testing are also considered.

2. Design Methodology

2.1. Specification of the Electric Motorcycle

In this paper, the SRM is designed for an electric motorcycle, comparable to a conventional internal combustion engine (ICE) with a motorcycle size of 120 cc. The targeted specifications of the electric motorcycle for one or two passengers, including maximum speed, acceleration from 0–30 km/h, and driving range per charge (at a constant speed of 40 km/h), are indicated in Table 1. This targeted specification is considered one of the suitable driving patterns in urban areas with heavy traffic, which requires good acceleration, a high payload, a long driving distance, and a moderate vehicle speed.

Table 1. Targeted specification of the electric motorcycle.

Parameter	Unit	1 Passenger	2 Passenger
Maximum speed	km/h	80	70
Acceleration 0–30 km/h	s	4	6
Driving Range at 40 km/h	km	90	80
Payload	kg	75	150

In order to simplify the mechanical works, the conventional engine that is directly connected to the existing continuous variable transmission (CVT) gearbox will be replaced by the designed SRM. The CVT amplifies and transfers torque to the rear wheel. Its gear ratio can vary from 2.159:1 to 0.8:1, depending on the input shaft rotational speed of the engine or motor. The centrifugal clutch begins to transmit torque to the rear wheel when the input shaft rotates at about 1200 rpm, so the CVT is one of the key mechanical parts taken into the calculation of the SRM's required torque and speed characteristics. Spaces below the footrest and inside the driver's seat are modified to install the inverter, the battery charger, and the 60 Ah 51.20 V Li-ion battery packs along with the battery management system (BMS). The overall system and the developed prototype are shown in Figure 1.

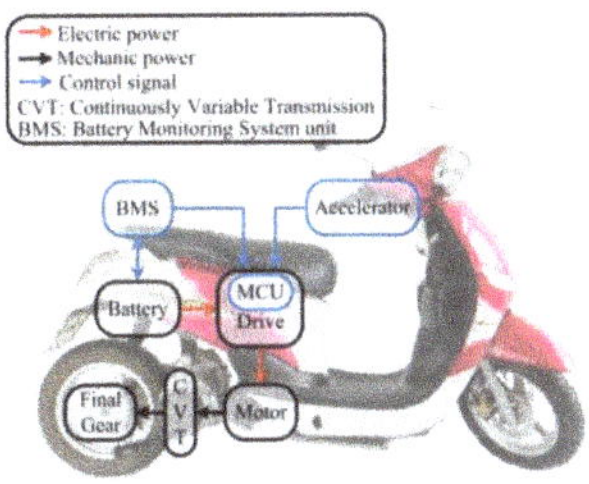

(a)

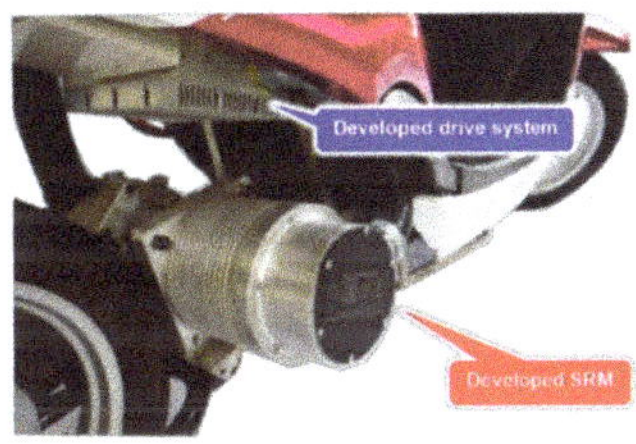

(b)

Figure 1. The electric motorcycle driven by the SRM and drive system. (**a**) The schematic of the electric motorcycle system; (**b**) the installed SRM and drive system.

The tractive force necessary to fulfill the requirements of vehicle performance in Table 1 can be calculated using vehicle dynamic equations from [15], such as the following key Equations:

$$F_T - (F_a + F_g + F_w) = C_m m a \quad (1)$$

where the tractive force F_T must exceed all resistance forces, including the aerodynamic drag F_a, the grade resistance due to gravity F_g, and the wheel resistance F_w. The mass of the vehicle is m, a is the acceleration, and C_m is the inertia coefficient that represents the inertia of the rotating mass, such as the gears and wheels. Each resistance force could be defined by

$$F_a = \frac{1}{2}\rho A_f C_D (V - V_w)^2 \quad (2)$$

$$F_g = mg \sin\alpha \quad (3)$$

$$F_w = f_r mg \cos\alpha \quad (4)$$

where ρ is the density of air, C_D is the drag coefficient of the vehicle depending mainly on its shape, A_f is the frontal area of the vehicle, V_w is the wind speed in the opposite direction of the vehicle's movement, and α is the grade angle. For a motorcycle, C_D is between 0.5 and 1.0.

The tractive force requirements can be calculated from (1)–(4) in the design process to achieve the desired driving performance, as shown in Table 1. Then, the traction motor torque T_{mot} can be calculated by

$$T_{mot} = \frac{F_T \, r_w}{G_r \times \eta_G} \quad (5)$$

where r_w is the tire radius, G_r is the gear ratio, and η_G is the gearbox efficiency.

The required torque speed characteristics of the traction system in the low-speed region include a requirement for high torque in order to shorten accelerating time, especially during hill climbing, as well as a wide power range in the high-speed region. From the above-mentioned equations and the dynamometer testing results of one commercial electric vehicle, which has a similar targeted specification, it can be seen that at least 8 Nm of torque in the low-speed region and about 7900 rpm of motor speed (which equals to a vehicle speed of 80 km/h) are required.

Figure 2 shows the overall electric motorcycle design process under consideration. The electric motorcycle specifications are cascading as subsequent targets for the following subsystem design. The specification of each subsystem, namely the traction motor, the motor drive system, battery system and mechanical structure, are taken into consideration since they have interactive impacts on the performance of the vehicle. Several iterative simulations have been conducted between system and subsystem levels in order to attain the desired design goals. These interactions include motor torque–speed characteristics, motor weight and size, DC voltage, battery DC current, drive and motor current, battery weight and size, and battery capacity. As the battery system is one of the key components in the electric motorcycle, the $LiFePO_4$ battery pack has been chosen for its high capacity per weight ratio, size, and charge/discharge rate.

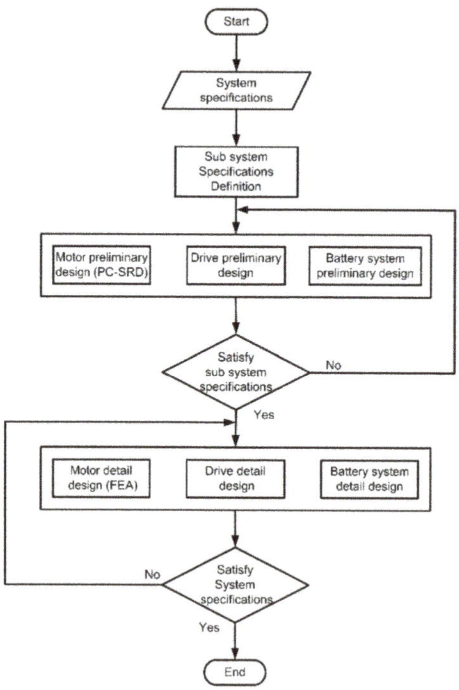

Figure 2. The flowchart of the electric motorcycle design process.

2.2. Traction Motor Design Consideration

The SRM was designed using a commercial analytical software package PC-SRD [16] and a 2D finite element analysis (FEA) using JMAG software. PC-SRD was used in the preliminary design phase to compare and adjust various design parameters, such as the number of stator/rotor poles, the number of phases, and the winding parameters. The software can calculate torque, current, efficiency, copper loss, iron loss, etc. It also includes a core structure, core material data, such as BH curves, iron loss curves, winding structure, applied voltage, rotational speed, and the on-off timing of an inverter. In order to gain high starting torque in the low-speed region, a four-phase 8/6 pole SRM and a three-phase 6/4 pole SRM were selected for performance comparison. Moreover, the weight and cost of the traction motor are also factors under consideration.

The design procedure of the switched reluctance machines starts with the selection of a frame size, which is performed in accordance with the existing standard of induction frame size (IEC71). The selection of frame size according to the IEC recommendations fixed the outer diameter of the stator and stack length in order to maintain a similar volume. According to the available space in the motorcycle, the key initial parameters included the stator's outer diameter, which was fixed to 140 mm, and its stack length, which was fixed to 70 mm, as shown in Table 2.

Table 2. Key initial parameters of each motor.

Parameter	Unit	8/6 Pole SRM	6/4 Pole SRM
Rated motor voltage (DC)	V	48	48
Outer stator dimension	mm	140	140
Stack length	mm	70	70
Air gap	mm	0.3	0.3

Based on the guideline for selecting a suitable stator and rotor pole arc in [17,18], a stator pole arc at β_s = 30 degrees and a rotor pole arc at β_r = 31.5 degrees were selected as the initial parameters for the three-phase 6/4 pole SRM. Additionally, the air gap was set to 0.30 mm. The silicon steel with a thickness of 0.35 mm (35A300) was used and its BH curve was used for analysis in the simulation. The magnetic characteristics of iron cores are particularly important because SRMs do not use a permanent magnet. In this paper, the 35A300 in JIS, i.e., a general low-loss silicon steel, was used. This steel has a thickness of 0.35 mm and is equal to 36F168, 300, and V300-35A in AISI, BS, and DIN, respectively.

The torque performance in the low-speed region (2000 rpm) of the four-phase 8/6 pole and three-phase 6/4 pole SRMs was investigated for each combination of rotor yoke, stator yoke, rotor pole width, stator pole width, as well as the number of turns and number of strands, in order to satisfy the requirements presented in Table 1. Each parameter was varied under the same boundaries of maximum phase current, maximum current density, and flux density. Then, the maximum value of the developed torque was determined (optimal torque operation). The calculated outline geometry and winding parameters of both SRMs are shown in Figure 3 and Table 3. The preliminary results of both SRMs at 2000 rpm are shown in Table 4. It was found that both motors could generate torque larger than 8 Nm and the three-phase 6/4 pole SRM has a 13.2% larger torque and a 14.7% larger torque per weight than the four-phase 8/6 pole SRM. The total cost of the copper and iron of the three-phase 6/4 pole SRM is 6.5% more expensive, and its efficiency is 9.8% lower than the 8/6 pole SRM.

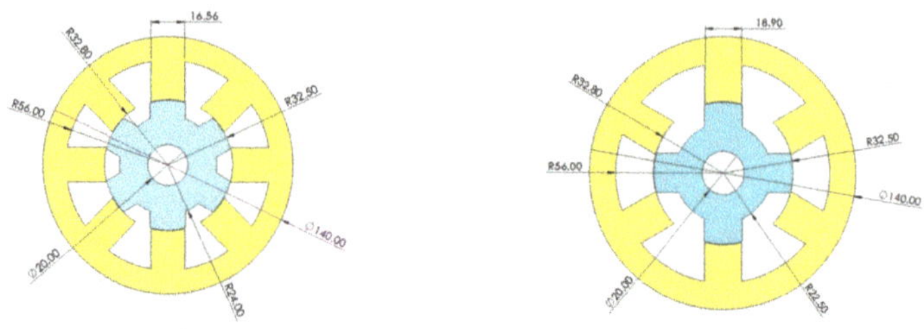

(a) (b)

Figure 3. Outline of the designed motor: (**a**) four-phase 8/6 pole SRM; (**b**) three-phase 6/4 pole SRM.

Table 3. Calculated parameter of each motor.

Parameter	Unit	8/6 Pole SRM	6/4 Pole SRM
Radius to the bottom of the rotor slot (R0)	mm	24	22.5
Rotor surface radius (R1)	mm	32.5	32.5
Radius to the bottom of the stator slot (R2)	mm	56	56
Stator outside radius (R3)	mm	70	70
Air gap	mm	0.3	0.3
Stator pole arc (β_s)	deg	22.5	24
Rotor pole arc (β_r)	deg	20	31.05
Number of turns	-	9	16
Number of strands	-	11	11
Wire diameter	mm	0.914	0.813
Material	-	35A300	35A300
Rotor inertia	kg*·m^2	4.8777×10^{-4}	5.1453×10^{-4}

Table 4. Preliminary result of the four-phase 8/6 pole and three-phase 6/4 pole SRM at 2000 rpm.

Parameter	Unit	8/6 Pole SRM	6/4 Pole SRM
Phase resistance	Ohm	0.012	0.032
Aligned inductance	mH	0.546	2.065
Unaligned inductance	mH	0.062	0.264
Theta on	deg	29	34
Theta off	deg	58	80
Average torque	Nm	8.34	9.44
Phase current (rms)	A	85.62	82.8
Shaft power	W	1746.37	1976.89
Overall system efficiency	%	80.46	73.27
Copper loss	W	362	659
Iron loss	W	55	53
Mechanical loss	W	6	8
Flux density at rotor pole	T	2.179	1.723
Flux density at stator pole	T	1.922	2.197
Copper weight	kg	1.3305	1.5959
Iron weight	kg	5.3165	4.9165
Torque per weight	Nm/kg	1.25	1.44
Total copper and iron cost	$	52.75	56.19

2.3. Static Performance Consideration

Figure 4 shows the torque, power, and efficiency characteristics of the designed four-phase 8/6 pole SRM and three-phase 6/4 pole SRM at each motor rotating speed. In the low-speed region, the motor is controlled by the current control mode, and it is controlled by voltage control in the high-speed region. At each operating speed, the turn-on and turn-off angles were adjusted under the same maximum value of the hysteresis band in order to search for the best combination value that serves either the optimal torque or optimal efficient operation. To find the best turn-on and turn-off control angles for the maximum torque and efficiency of the SRM, the grid search method, which is the easiest to implement for the small number of parameters, was applied to reach the optimal control angles. The optimization grid search method was implemented in Visual Basic for Applications (VBA) of Microsoft Excel, which can be linked to PC-SRD motor design software. Grid search was performed across different values of turn-on and turn-off angles. The different combinations of turn-on and turn-off angles are used as input for PC-SRD to evaluate and compare the torque and efficiency to find the optimal turn-on and turn-off angles.

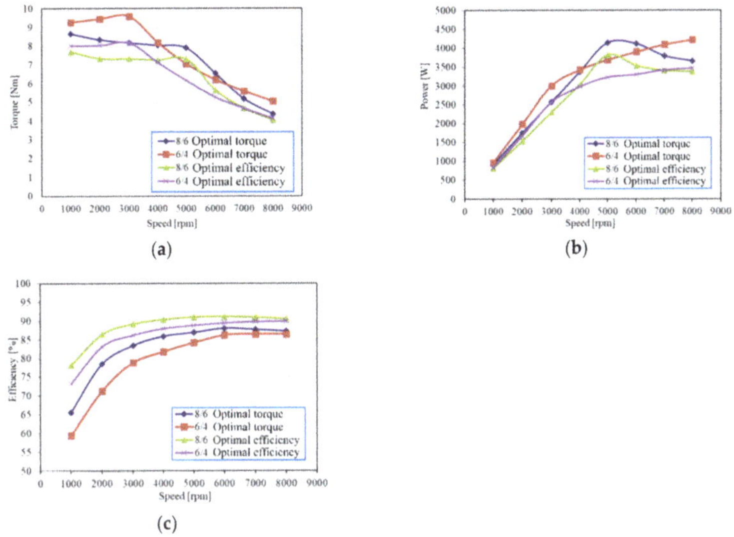

Figure 4. The simulated performance of the designed four-phase 8/6 and three-phase 6/4 SRM under optimal torque and optimal efficiency operations: (**a**) average torque versus rotating speed; (**b**) power versus rotating speed; (**c**) efficiency versus rotating speed.

Two objectives were defined for the optimization of this traction motor. The first objective is to search for the conduction angle that can give the highest torque at each speed. The second objective is to find the conduction angle that can achieve the highest efficiency at each speed. Constraints are that the torque of the second objective must be within the threshold level of 1.5 Nm of the required torque, maximum DC current is 120 A, and maximum DC voltage is 48 Vdc. Design variables are the turn-on angle (20–60 deg. for 6/4 SRM and 20–40 deg. for 8/6 SRM) and the turn-off angle (65–85 deg. for 6/4 SRM and 45–65 for 8/6 SRM). The objective function can be given in

$$Objective\ Function = \eta + k_{torque} \cdot Torque \tag{6}$$

where η and k_{torque} are the efficiency of the SRM and a factor which converts torque into an efficiency equivalent, respectively.

Equation (6) is maximized using a grid search optimization algorithm across the observed values of the threshold variable. Finally, the turn-on and turn-off angles for maximum torque and maximum efficiency at each speed (1000–8000 rpm) are obtained. For the optimal efficiency operation, the peak efficiency for the four-phase 8/6 pole SRM of about 91.13% at 6000 rpm can be achieved, which is not different from the peak efficiency for the three-phase 6/4 pole SRM of about 89.87% at 8000 rpm. Furthermore, for the optimal torque operation, in the low-speed region, the maximum average torque for the three-phase 6/4 pole SRM of about 9.57 Nm can be achieved, which is bigger than the maximum average torque for the four-phase 8/6 pole of about 8.66 Nm. That means the three-phase 6/4 pole SRM has better torque performance in the low-speed region and good efficiency performance in the high-speed region, which are preferable characteristics for electric motorcycle application.

The optimal angles for maximum torque were applied to control the vehicle during the low-speed regions or when a high dynamic response was required (braking and acceleration). The optimal angles for maximum efficiency were used to control the vehicle at medium to high-speed regions or at low torque requirements to increase the range of travel and reduce the power consumption of the battery.

2.4. Dynamic Performance Consideration

Basically, the three-phase 6/4 pole SRM has large torque ripple in the low-speed region. Therefore, it was necessary to observe the motor phase current and torque waveforms. The 2D FEA models for dynamic simulation are revealed in Figure 5. Figures 6 and 7 show the dynamic simulation at 2000 rpm under the optimal torque operation calculated by PC-SRD and FEA for the four-phase 8/6 pole and three-phase 6/4 pole SRM, respectively. The calculated values of the average torque and torque ripple rate are shown in Table 5. According to the FEA calculation, some effects from the magnetic coupling between the phases could be clearly observed on the phase current and torque waveforms. The results calculated by FEA are smaller than the results calculated by PC-SRD. The three-phase 6/4 pole SRM has a larger torque ripple rate than the four-phase 8/6 pole SRM. The effect of torque ripple would be absorbed by the CVT gearbox. However, the three-phase 6/4 pole SRM also has a larger average torque. Therefore, the three-phase 6/4 pole SRM was selected and the final design verification was conducted. Afterwards, FEA was used to finalize the design before constructing the prototype as shown.

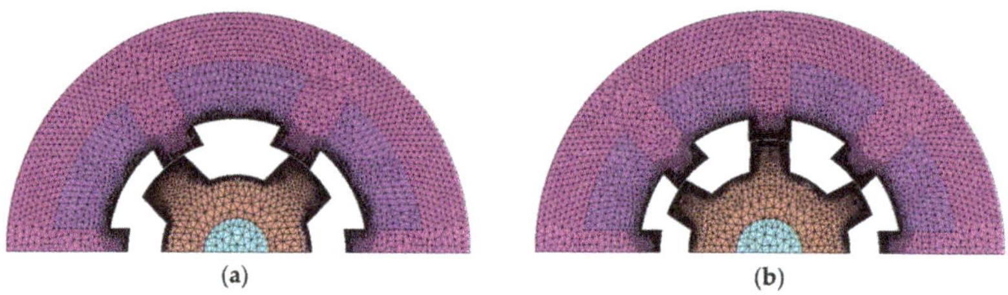

Figure 5. Mesh size of FEA model (**a**) 6/4 SRM; (**b**) 8/6 SRM.

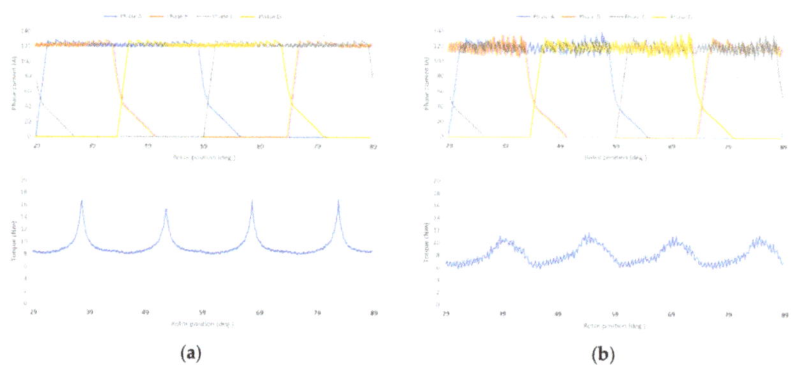

Figure 6. Four-phase current and total torque waveforms of the four-phase 8/6 pole SRM at 2000 rpm calculated from (**a**) PC-SRD; (**b**) FEA.

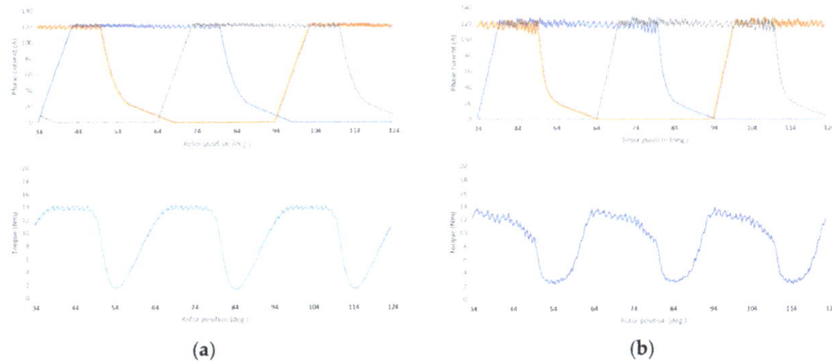

(a) (b)

Figure 7. Three-phase current and total torque waveforms of the three-phase 6/4 pole SRM at 2000 rpm calculated from (**a**) PC-SRD; (**b**) FEA.

Table 5. Calculated average torque and torque ripple rate of each motor.

	8/6 Pole SRM		6/4 Pole SRM	
	PC-SRD	FEA	PC-SRD	FEA
Average torque (Nm)	8.34	8.23	9.44	9.15
Torque ripple rate	0.96	0.72	1.33	1.27

The result of the preliminary three-phase 6/4 pole SRM design was verified by FEA, which can compute the magnetic field distribution with a ferromagnetic material saturation effect within the structure of the SRM. The results indicate that the flux density levels are within the limit of the chosen 35A300 silicon steel material. The flux linkages of SRM at different current levels are shown in Figure 8, where the rotor position is varied from an unaligned to aligned position. The electromagnetic torque was calculated for various current levels where the maximum instantaneous torque of 18 Nm was achieved at 180 A, as shown in Figure 9.

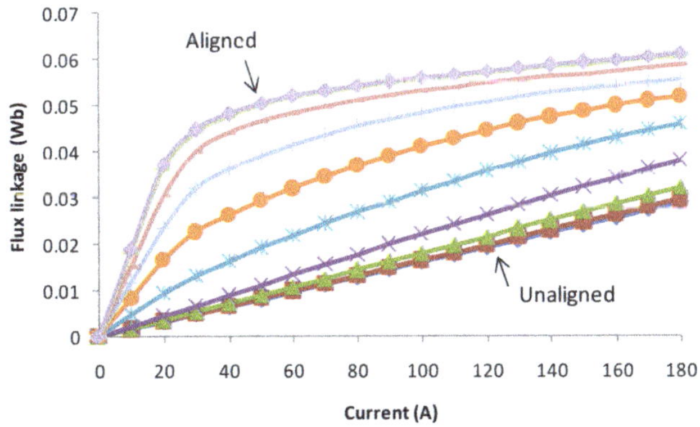

Figure 8. Flux linkage calculation from FEA.

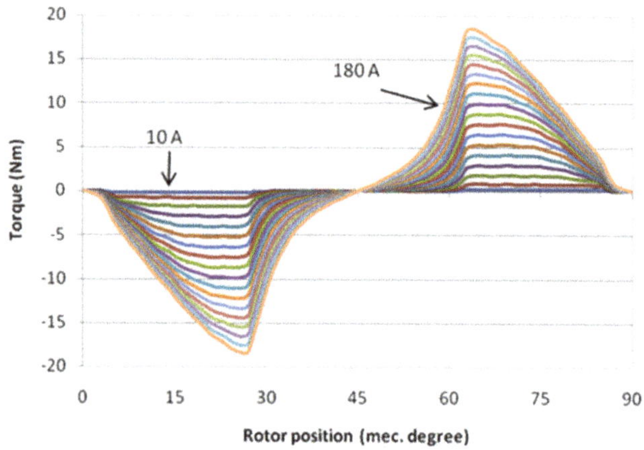

Figure 9. Torque versus current and rotor position.

Winding arrangement is another factor that could affect the static torque performance of the SRM. Figure 10 shows two winding configurations of the three-phase 6/4 pole SRM when a single phase (phase A) is excited with a unipolar operation. Figure 11 shows the flux density and flux path of the short-flux path and long-flux path winding configurations, respectively. The comparison of the static torque developed at each rotor position for both the short-flux and long-flux paths of the three-phase 6/4 pole SRM is based on the same conduction loss condition, which is also shown in Figure 12. The long-flux path winding could produce more positive torque during the effective torque production region, which is 45 mechanical degrees for the three-phase 6/4 pole SRM. The short-flux path winding could produce lower peak torque and it produces negative torque at 77–90 degrees. The three-phase 6/4 pole SRM should also be excited by bipolar excitation with a sinusoidal waveform [19] in order to reduce the torque ripple rate. However, the average torque and efficiency are also smaller than the unipolar excitation. Therefore, the static torque performance of the three-phase 6/4 pole SRM excited by bipolar excitation is not investigated here.

(a) (b)

Figure 10. Winding configuration of the three-phase 6/4 pole SRM: (**a**) short-flux path; (**b**) long-flux path.

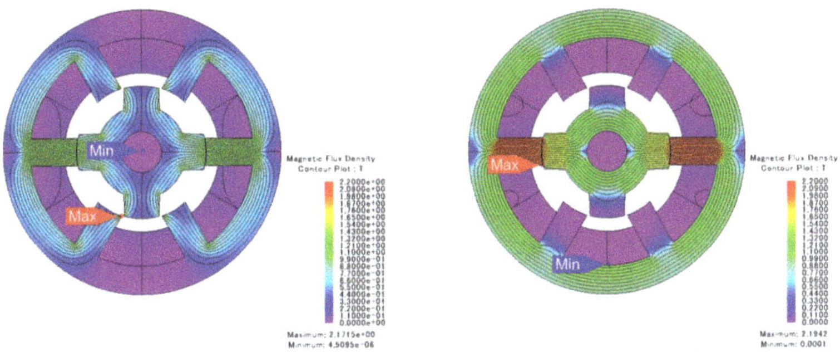

(a) (b)

Figure 11. Flux distribution of each winding configuration: (**a**) short-flux path; (**b**) long-flux path.

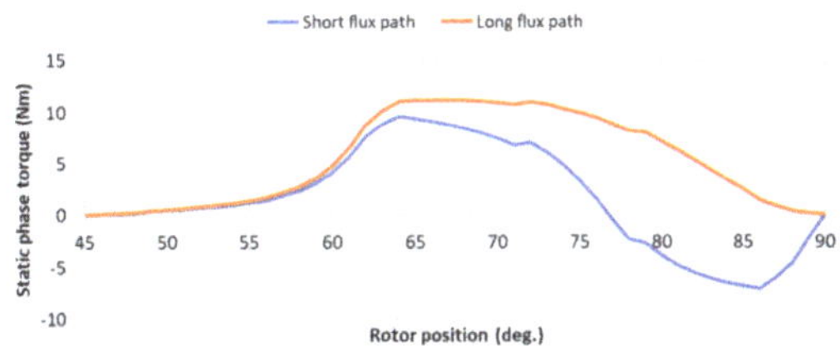

Figure 12. Static phase torque for each winding configuration.

3. Construction of the Motor and Drive System

Figure 13 shows the constructed three-phase 6/4 pole SRM prototype and drive system. In total, 30 units of the three-phase 6/4 pole SRM were manufactured by the private company in order to control the quality of the prototypes.

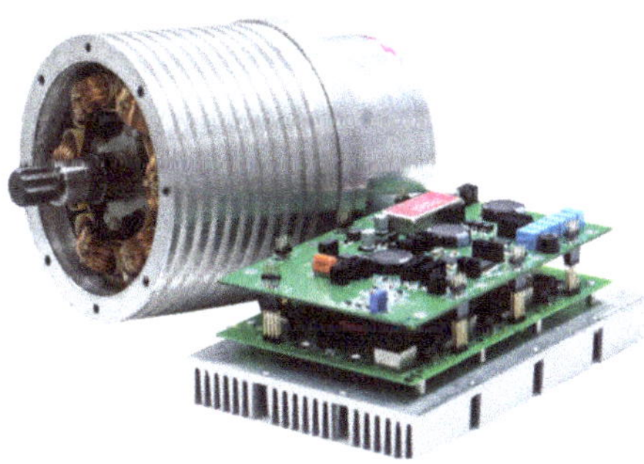

Figure 13. The prototypes of the three-phase 6/4 pole SRM and drive system.

The motor drive system employs the conventional three-phase, asymmetric half-bridge inverter. This topology has the advantages of simplicity, robustness, and fault tolerance, where each phase can be controlled independently with unidirectional current flow. These greatly enhance the safety features of the electric motorcycle drive system. The control block diagram for the SRM drive system is shown in Figure 14, where the user turns the accelerator to define the torque command as an input. The motor torque is estimated based on the torque to current block. The optimal turn-on angle, turn-off angle, and current command for maximum efficiency operation (optimal efficiency operation) at each torque and speed level will be calculated off-line by FEA and put into the 3D look-up table, as shown in Figure 15. At each torque command and detected motor speed, the corresponding current command and on/off angles will be calculated by the following Equation.

$$f(\tau_i, s_i) = \frac{s_i - s_2}{s_1 - s_2}\left[\frac{\tau_i - \tau_2}{\tau_1 - \tau_2} f(\tau_1, s_1) + \frac{\tau_i - \tau_1}{\tau_2 - \tau_1} f(\tau_2, s_1)\right] + \frac{s_i - s_1}{s_2 - s_1}\left[\frac{\tau_i - \tau_2}{\tau_1 - \tau_2} f(\tau_1, s_2) + \frac{\tau_i - \tau_1}{\tau_2 - \tau_1} f(\tau_2, s_2)\right] \quad (7)$$

where τ_i and s_i are the torque command and detected motor speed, respectively.

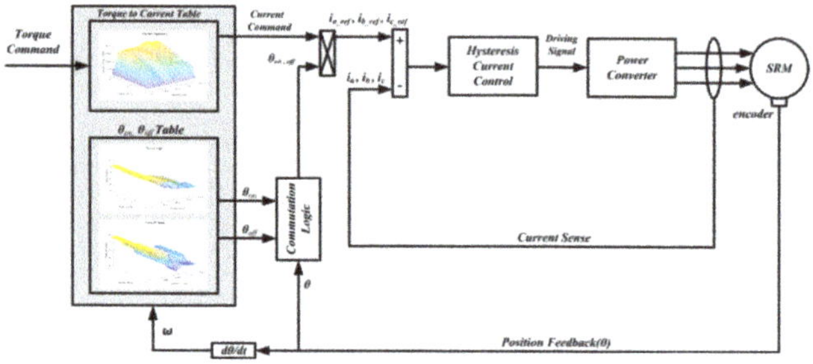

Figure 14. The control block diagram.

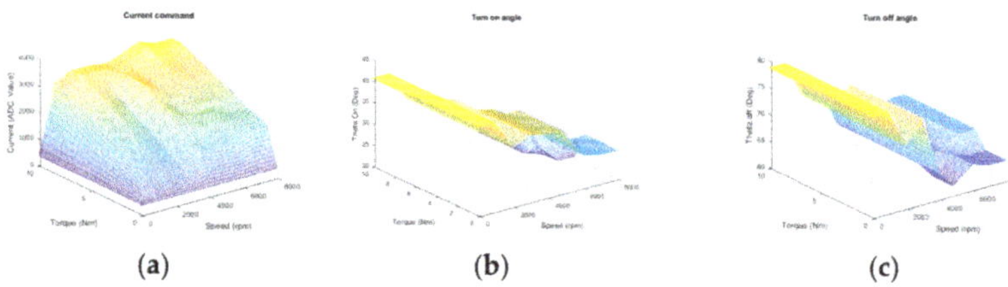

Figure 15. Optimal value of (**a**) current command; (**b**) turn-on angle; and (**c**) turn-off angle at each torque command and motor speed level for optimal efficiency operation.

The rotor position is also a necessary parameter for the indirect torque controller and current controller in the drive system. For the electric motorcycle application, especially when the vehicle with maximum payload starts from the standstill condition on the slope, the initial rotor position is very necessary for high starting torque development. Some low-cost, incremental optical encoders may be used for rotor position detection. However, they cannot detect the initial rotor position at a standstill condition, leading to torque jerk [20], which seriously affects the safety of the electric motorcycle. Furthermore, due to the harsh environment on the road, with dust, water, and vibration, a more reliable detection system for a non-contact rotor position is required. In this paper, the contactless magnetic-based position sensor was installed on the rotor shaft for position and speed feedback, as shown in Figure 16. It is a very low-cost and high-reliability solution. With the proposed mechanical coupling device using a bolted shaft and ball bearing, the effect of the deviated distance between the permanent magnet and sensing circuit caused by speed variation and vibration could be eliminated effectively. A resolution of the rotor position detection of about 0.87 degrees could be achieved from 0 to 8000 rpm. The proposed initial rotor position estimation method [20] based on pulse injection and phase current profile has a maximum error of about 2.5 degrees, but it could identify the sector of the rotor position. Therefore, the combination of the proposed rotor position estimation and magnetic sensing system could increase the reliability of the electric motorcycle.

(a) (b)

Figure 16. Rotor position detection system: (**a**) mechanical coupling device; (**b**) a constructed rotor position detection system installed on the three-phase 6/4 pole SRM.

The motor phase current is sensed by current sensor model ACS758ECB-200U which can be mounted on a PCB. The current regulation is also performed by an inner digital hysteresis controller. A 32-bit micro-controller, STM32F103RET6, and a power MOSFET, IRF4468, were used. Figure 17 shows the battery pack and circuit board in the battery management system. In one battery pack, a total of eight cells of LiFePO4 model NLC36130185PF (3.2 V 63 Ah) were connected in a series and two sets of battery packs were installed in each electric motorcycle. The voltage of each cell is monitored by the cell monitoring board and it is transmitted to the master board. The master board will perform passive balancing by bypassing the current to the cell that has the lower capacity and it will make a decision for communication to the quick charger and interlock circuit for safe operation.

(a) (b) (c)

Figure 17. (**a**) Lithium ion battery pack; (**b**) master board of the battery management system; (**c**) cell monitoring board of the battery management system.

4. Experiment Results

4.1. Motor and Drive System Test Result

The static test of the SRM prototype was conducted. The electrical characteristics of the prototype, such as the resistance and inductance, can be measured via a static test with the locked rotor. Table 6 shows the measured inductance and simulated value obtained from FEA. The small discrepancy may be caused by the effect of 2D FEA.

Table 6. Measured inductance value compared to FEA.

Phase	L-Aligned (mH)	L-Unaligned (mH)	L-Ratio
A	1.9767	0.2960	6.6774
B	1.8016	0.3342	5.3904
C	1.9385	0.2864	6.7666
FEA	2.0652	0.2643	7.8150

Figure 18 shows the efficiency map of the SRM prototype from the test results on the test bench. With the limited capacity of the testing equipment, the SRM drive was supplied with a 50 V_{DC} voltage source and maximum DC current of about 100 A, which is less than the value used in the simulation. Therefore, performance of up to 7000 rpm was measured.

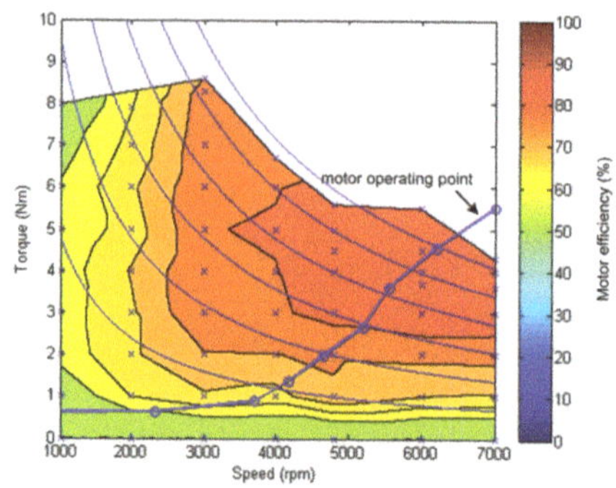

Figure 18. Efficiency map of the constructed three-phase 6/4 pole SRM.

The maximum torque measured via a torque transducer was approximately 8.6 Nm at 3000 rpm. The motor has an efficiency of more than 80% at a high speed and medium torque output range (81.85% at 7000 rpm), which is very suitable for an electric motorcycle running at medium to high speed. This load condition is the most practical scenario for the normal usage of a general purpose electric motorcycle.

4.2. Electric Motorcycle Performance Test Result

The performance of the constructed electric motorcycle was tested by a chassis dynamometer, as shown in Figure 19. The driving range of the electric motorcycle at a constant 40 km/h cruising speed (motor speed 4686 rpm) with a 150 kg pay load was measured. Figure 20 shows the measured values during the test run. From the starting time to 140 min, the measured current was about 30 A and the measured voltage gradually decreased from 51.20 V. When the running time reached 140 min, the battery voltage was lower than the battery threshold, and the drive system was then turned off. The total driving range of about 98 km could be achieved. The maximum vehicle speed and acceleration from 0 to 30 km/h during on-road testing with full payload (150 kg) were measured and compared to the commercial electric motorcycle, as shown in Figure 21. The maximum speed could reach 82 km/h and the acceleration time was about 5.1 s, which is better than the requirements in Table 1.

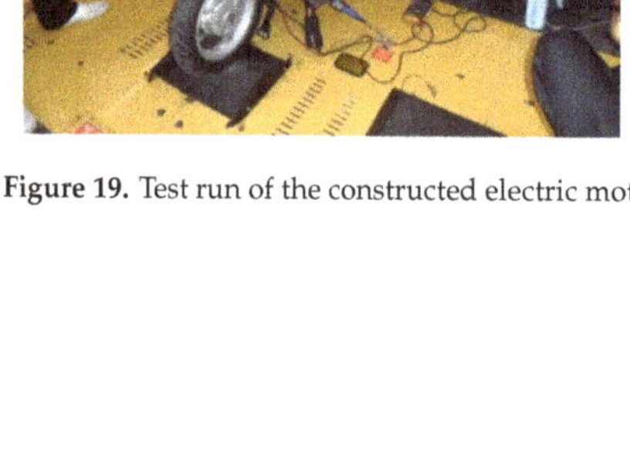

Figure 19. Test run of the constructed electric motorcycle on a chassis dynamometer.

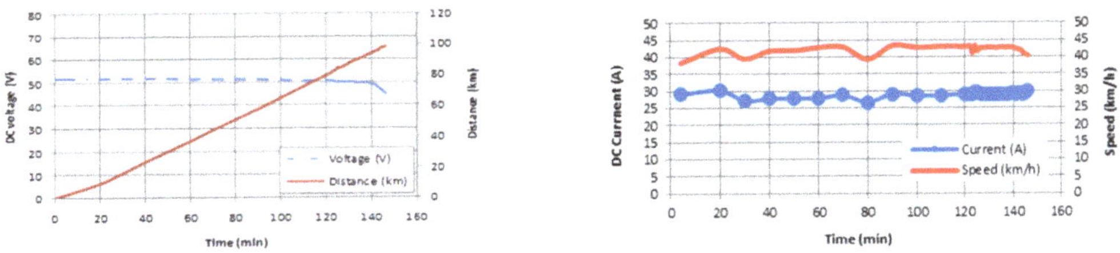

(a) (b)

Figure 20. (**a**) Measured DC voltage and distance; (**b**) measured DC current and vehicle speed.

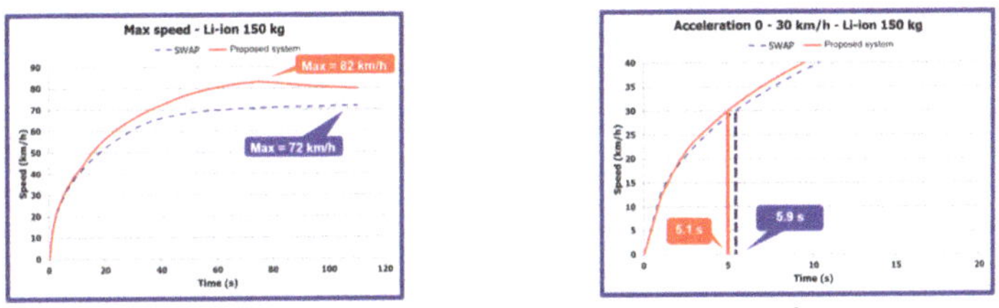

(a) (b)

Figure 21. (**a**) Maximum vehicle speed; (**b**) acceleration from 0 to 30 km/h.

The constructed electric motorcycle was tested and complied with the electromagnetic compatibility standards, including radiated emission and immunity testing based on CISPR25:2002 and ISO 11452-2, respectively, as shown in Figure 22. Furthermore, the SRM drive system was also in compliance with the IEC1-60335 standard for the safety of household and similar electrical appliances. The SRM was also tested and complied with the IEC60529:2001 (IP56) standard. All 30 units of the completed electric motorcycles were produced and transferred to the Provincial Electricity Authority (PEA) for their internal routine services.

Figure 22. EMI/EMC testing of the constructed electric motorcycle.

5. Conclusions

This paper described the performance improvement of an electric motorcycle driven by an SRM and drive system. The static performance comparison of the three-phase 6/4 pole and four-phase 8/6 pole SRM was undertaken. It was found that the three-phase 6/4 pole SRM had a larger starting torque and torque per weight in the low-speed region. In the high-speed region, both SRMs could reach about 90% with the optimal efficiency operation. The average and torque ripple rate of both SRMs were also investigated by dynamic simulation. The torque ripple rate of the three-phase 6/4 pole SRM was a bit larger than the four-phase 8/6 pole SRM. However, the three-phase 6/4 pole SRM also had a larger average torque. Therefore, the three-phase 6/4 pole SRM was also suitable for the propulsion system of the electric motorcycle. The performance of the three-phase 6/4 SRM was also investigated by FEA. The implementation of indirect torque control and rotor position detection in the drive system was proposed. According to the testing results obtained on the test bench, the highest motor efficiency of about 82% could be achieved. The vehicle performance, including a maximum speed of 82 km/h and a driving range of 98 km at a constant speed condition of 40 km/h, could be also achieved.

Author Contributions: Conceptualization, S.K., P.J., P.S., R.P. and K.T.; methodology, S.K., P.J., P.S. and K.T.; validation, R.P., S.K., P.J., P.S. and K.T.; formal analysis, S.K., P.J., P.S., R.P. and K.T.; investigation, R.P., P.S. and K.T.; resources, S.K., P.J. and P.S.; writing—original draft preparation, K.T.; writing—review and editing, R.P., S.K., P.J., P.S. and K.T.; visualization, R.P. and S.K.; supervision, P.S. and K.T. All authors have read and agreed to the published version of the manuscript.

Funding: This research was funded by the Provincial Electricity Authority Research Fund of Thailand.

Institutional Review Board Statement: Not applicable.

Informed Consent Statement: Not applicable.

Data Availability Statement: Not applicable.

Conflicts of Interest: The authors declare no conflict of interest.

References

1. Kerdlap, P.; Gheewala, S.H. Electric motorcycles in Thailand—A life cycle perspective. *J. Ind. Ecol.* **2016**, *20*, 1399–1411. [CrossRef]
2. Bostanci, E.; Moallem, M.; Parsapour, A.; Fahimi, B. Opportunities and challenges of switched reluctance motor drives for electric propulsion: A comparative study. *IEEE Trans. Transp. Electrif.* **2017**, *3*, 58–75. [CrossRef]
3. Rahman, K.M.; Fahimi, B.; Suresh, G.; Rajarathnam, A.V.; Ehsani, M. Advantages of switched reluctance motor applications to EV and HEV: Design and control issues. *IEEE Tran. Ind. Appl.* **2000**, *36*, 111–121. [CrossRef]
4. Lan, Y.; Benomar, Y.; Deepak, K.; Aksoz, A.; Baghdadi, M.; Bostanci, E.; Hegazy, O. Switched reluctance motors and drive systems for electric vehicle powertrains: State of the art analysis and future trends. *Energies* **2021**, *14*, 2079. [CrossRef]
5. Jiang, J.W.; Bilgin, B.; Emadi, A. Three-phase 24/16 Switched Reluctance Machines for a Hybrid Electric Powertrain. *IEEE Trans. Transp. Electrif.* **2017**, *3*, 76–85. [CrossRef]
6. Kiyota, K.; Chiba, A. Design of switched reluctance motor competitive to 60-kW IPMSM in third-generation hybrid electric vehicle. *IEEE Trans. Ind. Appl.* **2012**, *48*, 2303–2309. [CrossRef]
7. Wang, S.; Zhan, Q.; Ma, Z.; Zhou, L. Implementation of a 50-kW four-phase switched reluctance motor drive system for hybrid electric vehicle. *IEEE Trans. Magn.* **2005**, *41*, 501–504. [CrossRef]
8. Lin, J.; Schofield, N.; Emadi, A. External rotor 6-10 switched reluctance motor for an electric bicycle. *IEEE Trans. Transp. Electrif.* **2015**, *1*, 348–356. [CrossRef]
9. Terzic, M.V.; Bilgin, B.; Emadi, A. Switched reluctance motor design for a forklift traction application. In Proceedings of the 2018 XIII International Conference on Electrical Machines (ICEM), Alexandroupoli, Greece, 3–6 September 2018; pp. 812–818.
10. Andrada, P.; Blanque, B.; Capo, M.; Gross, G.; Montesinos, D. Switched reluctance motor for light electric vehicles. In Proceedings of the 2018 20th European Conference on Power Electronics and Applications (EPE'18 ECCE Europe), Riga, Latvia, 17–21 September 2018; pp. 1383–1388.
11. Vosswinkel, M.; Lohner, A.; Platte, V.; Hirche, T. Design, production, and verification of a switched-reluctance wheel hub drive train for battery electric vehicles. *World Electr. Veh. J.* **2019**, *10*, 82. [CrossRef]
12. Lin, C.H. Switched reluctance motor drive for electric motorcycle using HFNN controller. In Proceedings of the 2007 7th International Conference on Power Electronics and Drive Systems, Bangkok, Thailand, 27–30 November 2007; pp. 1383–1388.
13. Tomczewski, K.; Wrobel, K.; Rataj, D.; Trzmiel, G. A switched reluctance motor drive controller based on an FPGA device with a complex PID regulator. *Energies* **2021**, *14*, 1423. [CrossRef]
14. Anvari, B.; Toliyat, H.A.; Fahimi, B. Simultaneous optimization of geometric and firing angles for in-wheel switched reluctance motor drive. *IEEE Trans. Transp. Electrif.* **2018**, *4*, 322–329. [CrossRef]
15. Ehsani, M.; Gao, Y.; Longo, S.; Ebrahimi, K.M. *Modern Electric, Hybrid Electric, and Fuel Cell Vehicles Fundamentals, Theory, and Design*, 2nd ed.; CRC Press: Boca Raton, FL, USA, 2010.
16. SPEED Consortium University of Glasgow. *User's Manual: PC-SRD*; Version 8.8, CD-Adapco; SPEED Consortium University of Glasgow: Glasgow, UK, 2011.
17. Vijayraphavan, P. Design of Switched Reluctance Motors and Development of a Universal Controller for Switched Reluctance and Permanent Magnet Brushless DC Motor Drives. Ph.D. Thesis, Virginia Polytechnic Institute and State University, Blacksburg, VA, USA, 2001.
18. Miller, T.J.E. *Switched Reluctance Motors and Their Control*; Magna Physics Publishing and Clarendon Press: Oxford, UK, 1993.

19. Liu, X.; Zhu, Z.Q.; Hasegawa, M.; Pride, A.; Deodhar, R.; Maruyama, T.; Chen, Z. Performance comparison between unipolar and bipolar excitations in switched reluctance machine with sinusoidal and rectangular waveforms. In Proceedings of the 2011 IEEE Energy Conversion Congress and Exposition, Phoenix, AZ, USA, 17–22 September 2011; pp. 1590–1595. [CrossRef]
20. Tungpimolrut, K.; Jitkreeyarn, P.; Kachapornkul, S.; Somsiri, P.; Chiba, A. Initial rotor position estimation of a SRM drive installed in an electric vehicle. *IEEJ Trans. Electr. Electron. Eng.* **2011**, *6*, 594–600. [CrossRef]

energies

MDPI

Article

Design and Thermal Analysis of Linear Hybrid Excited Flux Switching Machine Using Ferrite Magnets

Himayat Ullah Jan [1,*], Faisal Khan [1], Basharat Ullah[1,*], Muhammad Qasim [1], Ahmad H. Milyani [2,3] and Abdullah Ahmed Azhari [4]

1 Department of Electrical and Computer Engineering, COMSATS University Islamabad, Abbottabad 22060, Pakistan; faisalkhan@cuiatd.edu.pk (F.K.); qasimrajput217@gmail.com (M.Q.)
2 Department of Electrical and Computer Engineering, King Abdulaziz University, Jeddah 21589, Saudi Arabia; ahmilyani@kau.edu.sa
3 Center of Research Excellence in Renewable Energy and Power Systems, King Abdulaziz University, Jeddah 21589, Saudi Arabia
4 Computer Information Technology, The Applied College, King Abdulaziz University, Jeddah 21589, Saudi Arabia; aaazhari1@kau.edu.sa
* Correspondence: jhimayatullah@gmail.com (H.U.J.); basharat.bigb@gmail.com (B.U.)

Abstract: This paper presents a novel linear hybrid excited flux switching permanent magnet machine (LHEFSPMM) with a crooked tooth modular stator. Conventional stators are made up of a pure iron core, which results in high manufacturing costs and increased iron core losses. Using a modular stator lowers the iron volume by up to 18% compared to a conventional stator, which minimizes the core losses and reduces the machine's overall cost. A crooked angle is introduced to improve the flux linkage between the stator pole and the mover slot. Ferrite magnets are used with parallel magnetization to reduce the cost of the machine. Two-dimensional FEA is performed to analyze and evaluate various performance parameters of the proposed machine. Geometric optimization is used to optimize the split ratio (S.R) and winding slot area ($Slot_{area}$). Genetic algorithm (GA) is applied and is used to optimize stator tooth width (S_{TW}), space between the modules (SS), crooked angle (α), and starting angle (θ). The proposed model has a high thrust density (306.61 kN/m^3), lower detent force (8.4 N), and a simpler design with higher efficiency (86%). The linear modular structure makes it a good candidate for railway transportation and electric trains. Thermal analysis of the machine is performed by FEA and then the results are validated by an LPMEC model. Overall, a very good agreement is observed between both the analyses, and relative percentage error of less than 3% is achieved, which is considerable since the FEA is in 3D while 2D temperature flow is considered in the LPMEC model.

Keywords: linear machine; flux switching machine; modular stator; crooked tooth; ferrite magnet; genetic algorithm; thermal analysis; LPMEC model

Citation: Jan, H.U.; Khan, F.; Ullah, B.; Qasim, M.; Milyani, A.H.; Azhari, A.A. Design and Thermal Analysis of Linear Hybrid Excited Flux Switching Machine Using Ferrite Magnets. *Energies* **2022**, *15*, 5275. https://doi.org/10.3390/en15145275

Academic Editors: Vladimir Prakht, Mohamed N. Ibrahim and Vadim Kazakbaev

Received: 30 June 2022
Accepted: 17 July 2022
Published: 21 July 2022

1. Introduction

Increase in the industrialization of the modern world increases the pollution caused by automobiles and other transportation sources. The modern world is shifting toward more efficient and pollution-free hybrid electric vehicles because of the environmental issues and to escape an imminent threat of energy scarcity. Railways also are the main constituent of the transportation system. Major intercity transportation is carried out by trains. Rotary motors have been used previously in the literature for the rail infrastructure, but the gearing system used to convert rotary motion to linear lowers the overall efficiency of the system. The trend is now shifting toward linear motors as they can be directly used because of their direct nonadhesive thrust force without using the gearing system.

Linear induction motors (LIMs) have been used recently for the transportation system. Primarily, the LIMs need a lower value of volume than most of the traditional motors, reducing the cross-sectional area of tunnels. A single-sided LIM is analyzed and studied in [1] with the derivation of the equivalent circuit and its analysis. Control strategies were devised for the LIMs in [2,3]. A modified model-predictive control was proposed for LIMs, and the results were analyzed. Effects of different secondaries were studied and investigated in [4,5]. Considering all these design optimizations and control algorithms, the main problems faced by LIMs are having lower efficiency and low power factor with respect to eddy currents, copper losses, and effects of edging, resulting in a high system and maintenance cost.

In contrast, linear permanent magnet (LPM) motors have been proven to have high efficiency, high power density, and high power factor. LPMs have many advantages, but they still have the drawback of using a large number of magnets. Mostly, the magnets are placed on the long stator, which not only makes it complex but also makes it costly. The linear flux switching machine (LFSM) is extensively studied nowadays because of its PM placement in the short mover and having a simple iron stator. In addition to these advantages of PM machines, LFSPM machines have several other advantages, such as easy maintenance, easy heat management system, and lower cost of the secondary, which makes it unique for purposes such as trains and railway stations [6,7]. However, LFSPM suffers from high detent force due to slot and end effects. Slot effect can be suppressed by adjusting the length and width of the PM [8]. It can be used to reduce detent force, but it has a negative effect on the thrust force, reducing it considerably. In [9,10], staggered tooth and semi-closed slots were proposed for reducing detent force and end effect, but it has the drawback of making the winding arrangement difficult and making the whole machine complex. To reduce the end effect in linear machines, auxiliary poles (APs) were proposed in [11,12], but they also increase the total harmonic distortions (THDs) of the back-EMF profile. Modular structure was used in [13,14] to reduce the end effect caused by unstable magnetic circuits.

Thermal environment greatly affects the working conditions of a machine. Authors in [15] used the Arrhenius model along with a coaxial multi-slot antenna. Authors in [16] analyzed various deformations along with the wear and tear because of the temperature rise. Ref. [17] only considered the distribution of temperature while ignoring the temperature rise with the passage of time. Refs. [18,19] divided the whole machine into various different parts and analyzed temperature rise in each of the parts, which increased the efficiency of the method. Authors in [20,21] used the same method of dividing the machine various isotropic parts and analyzing the temperature in each part and then comparing the average temperature rise in the whole of the machine.

This paper proposes a novel LHFSPM machine with crooked tooth modular stator. Two DC excitation sources are placed below and above a ferrite magnet, and overlapped concentrated winding is used for the armature. The stator of the machine is in the form of a U-shaped module with a crooked angle. The whole model is designed and analyzed in JMAG, registered version 20.1. No-load and loaded studies of the machine are carried out, and optimization techniques were used to improve the thrust force characteristics. This paper is divided into six further sections: Section 2 discusses the structure, design, and working principle of the machine; Section 3 covers optimization techniques and their effect on the performance of the machine; Section 4 discusses the electromagnetic performance evaluation using a number of study techniques to reach the proposed model; Section 5 discusses the thermal analysis of the machine and comparison between the FEA thermal analysis and LPMEC model; Section 6 presents the comparison of proposed and conventional machine; Section 7 provides the conclusion, which is an overview of the whole paper, and all the necessary points are discussed.

2. Design and Working Principle

The design of the proposed model is shown in Figure 1. DC field windings (shown by cyan color) and armature windings (shown by green color) are placed on the short mover. Permanent magnets (PMs) (shown by magenta color) are placed inset in the mover poles magnetized in a parallel direction. The stator of the model is in the form of U-shaped modules, which are placed at a certain distance from each other, having no direct electrical or magnetic contact. Such an arrangement uses less iron than the conventional stator and not only decreases the cost of the machine but also improves the overall efficiency of the machine. A conventional stator is made up of a full-length iron core which becomes impractical for countries such as Pakistan and other developing countries.

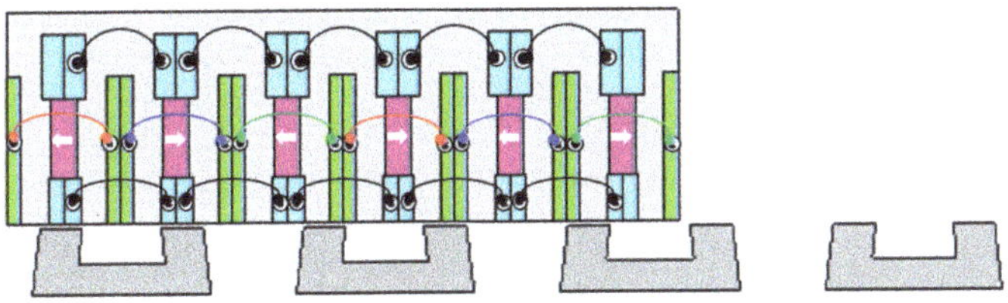

Figure 1. 2D view of the proposed model.

To minimize the detent force of the machine, a suitable selection of mover slot M_s and stator pole S_p is made using Equation (1) [13].

$$S_p = M_s\left(\frac{1 \pm n}{2q}\right) \quad (1)$$

where q denotes the number of phases and n represents any natural number. For various values of n, different stator pole numbers are studied and analyzed. It was noted that the machine has a sinusoidal flux linkage and unidirectional thrust force when $M_s/S_p = 6/5, 6/7$. The average thrust force for 6/5 is higher than 6/7, so it is considered for further analysis. The velocity of the machine is dependent on the input source frequency and pole pitch of the machine and can be found by using Equation (2) [22].

$$v = f\tau_s \quad (2)$$

where v is the velocity, f is the frequency of the source, and τ_s is the pole pitch of the machine. Flux linkage of the machine greatly varies (periodically) with the position of the mover relative to the stator position since the flux path changes as the mover changes its position. Considering phase A at two different points, both maximum and minimum flux linkage are analyzed, as shown in Figure 2a,b, respectively. At point 1, flux linkage due to phase A is at a positive maximum as the mover pole and stator tooth is completely aligned. Flux flows through the magnet, mover pole, air gap, and then into the stator tooth, completing the flux path back into the mover. A point 2, the flux linkage of phase A is at a negative maximum as the mover pole stator tooth is completely misaligned, while phase B and phase C have some value. The three phases are completely (120 degrees) apart. The direction of PM magnetization is set horizontal to that of the primary moving direction. The direction of PM and armature winding and DC is either clockwise or anticlockwise, to strengthen the overall machine's flux linkage. Parameters of the machine are defined in Figure 3. The details of the parameters and their numerical values are tabularized in Table 1.

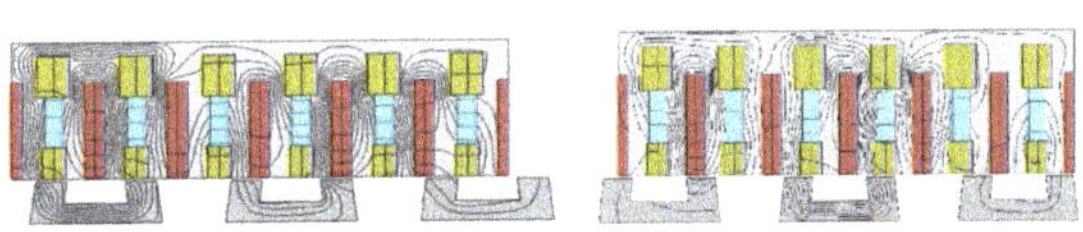

(a) (b)

Figure 2. Flux linkage at two different positions. (**a**) Point 1; (**b**) Point 2.

Table 1. Detailed parameters of the proposed model.

Parameters	Values
Pole Pitch, τ_s	26.1984 mm
Velocity, V	4 m/s
Stator Height, H_s	12.56 mm
Stack Length	90 mm
Mover Height, H_m	12.56 mm
Stator Tooth Width, S_{TW}	8.33338 mm
Air gap, L_g	0.8 mm
Armature winding turns	136
Maximum current DC	5.76 A
Permanent magnet	Ferrite_Br = 0.4 T
Maximum thrust force	150.1823 N
Split ratio, $S.R$	0.246
Armature slot width, A_w	2.4 mm
Armature slot length, A_L	31.241 mm
Upper DC slot width, DC_{UW}	12.55 mm
Upper DC slot length, DC_{UL}	4.3 mm
Lower DC slot width, DC_{LW}	3.119 mm
Lower DC slot length, DC_{LL}	9.264 mm

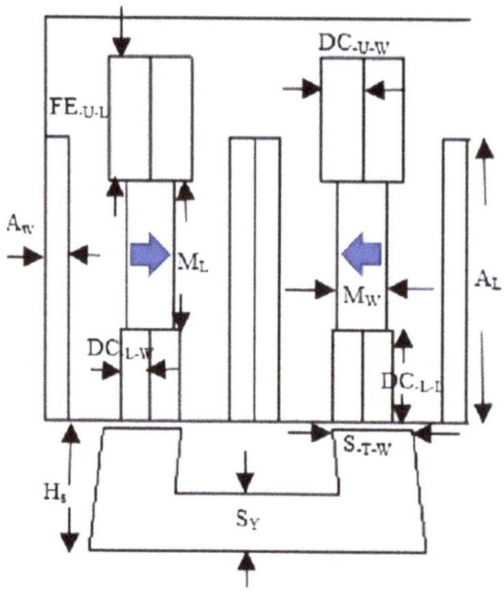

Figure 3. Parameters of the proposed model.

A flux density nephogram of the machine is shown in Figure 4, which reflects the magnetic flux density at various points of the machine, the path followed by the flux, and how flux switches from one mover pole to another pole through the modular stator. Slight saturation can be seen in the red regions in the nephogram, but it cannot be considered to cause a heating effect in the machine. The maximum flux density in the mover is 1.8 T, while in the case of the stator, maximum flux density is 1.42 T.

Coil configuration is dependent on the number of mover slots. Once the mover slot combination is confirmed, the coil configuration can be adjusted accordingly. Coil span is the axial distance through which a coil is wound. The coil span of the machine depends on the type of winding configuration used, either concentrated or simple.

$$C_s = \begin{bmatrix} max[fix[\frac{M_s}{T_s}]] & 1 \\ for\, concentrated\, winding & 1 \end{bmatrix} \quad (3)$$

where M_s represents number of mover slot, S_p represents number of stator tooth, and fix function is used to return integer value only. In case of a three-phase balanced system, the phases are separated 120 degrees apart. The phase separation for a machine can be found by C_0.

$$C_0 = \frac{N_m}{3N_s}(1+3k) \quad (4)$$

k is any integer value, e.g., $0, 1, 2, 3, \ldots \frac{N_s}{2}$. If a suitable value for k is not found, then that value of slot and pole combination shall not be chosen.

Figure 4. Nephogram of machine flux density.

3. Optimization and Refinement of Machine Parameters

To increase the thrust force and efficiency of the machine, different machine parameters were optimized using single variable geometric optimization and JMAG inbuilt optimization (genetic algorithm (GA)). Thrust force amplification is considered the main target. In geometric optimization, a series of consecutive values are considered for geometry, and then the resulting thrust force is analyzed.

3.1. Geometric Optimization

Leading parameters such as split ratio ($S.R$) and armature slot width (A_w) were optimized using geometric optimization. Figure 5 shows the flowchart of single variable geometric optimization.

Determination of optimal split ratio is a very important process of designing a machine as it decides not only the average thrust force but also the overall cost of the machine. If the selected value of the split ratio is low, it will reduce the stator height, which is suitable for railway transits. On the other hand, it will increase mover height and mass, resulting in the reduction of the average thrust force. The higher value of the split ratio resolves the mover mass problem but increases the stator volume, making the machine costly. The split ratio of the machine can be found by Equation (5) [23]. Table 2 shows the performance of the machine at different values of $S.R$.

$$S.R = \frac{H_s + L_g}{H_s + H_m + L_g} \quad (5)$$

$$A_w = \frac{Slot_{area}}{A_L} \quad (6)$$

Table 2. Performance indicator.

S.R	0.15	0.225	0.235	0.246	0.257
$TF_{average}$ (N)	119.17	121.72	128.43	133.48	129.13
$\varphi_{(p-p)}$ (Wb)	0.010	0.0107	0.017	0.012	0.018
$Fd_{(p-p)}$ (N)	0.0237	0.215	0.281	0.301	0.260
TF_d (kN/m^3)	243.30	248.51	262.21	272.52	263.64
THD (%)	12.93	11.23	13.75	14.1	13.42

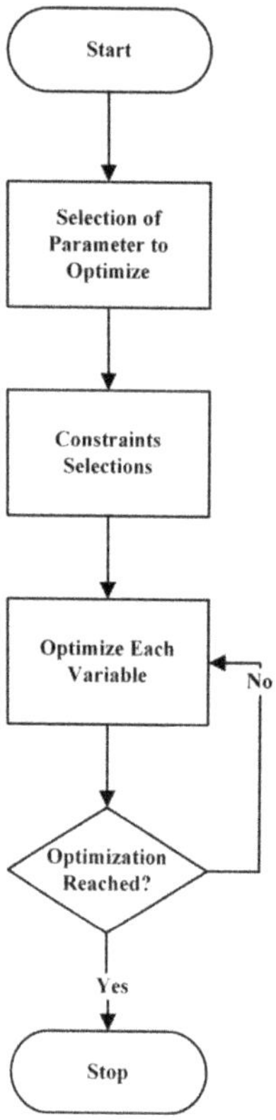

Figure 5. Flowchart of geometric optimization.

The suitable height and width of the slot area are selected by keeping the overall slot area constant and changing the width and height of the machine. Increasing the width tends to decrease the height of the slot, and decreasing the width increases the height of the slot. The height and width of the machine are interrelated by Equation (6). Table 3 shows the thrust force profile at different slot width and height values.

Table 3. Performance after geometric optimization.

Indicator	Stage1	Stage2	Stage3	Stage4	Stage5	Stage6
A_L (mm)	35.71	34.09	32.60	31.25	28.84	27.77
A_w (mm)	2.1	2.2	2.3	2.4	2.5	2.6
$Slot_{area}$ (mm^2)	75	75	75	75	75	75
$TF_{average}$ (N)	131.21	132.8	133.48	134.17	135.01	135.98

3.2. Genetic Optimization

Stator tooth width, stator module spacing, crooked angle of the tooth, and starting angle of the machine are the four parameters that are optimized using genetic optimization. The width of the tooth helps with better flux linkage and better alignment of the mover pole and stator tooth. Teeth provide the necessary path for the flux linkage, and if teeth are too thin, the machine will suffer from saturation, and if the teeth are too thick, the flux will not switch into the next mover pole. GA was used to select a better-suited stator width, resulting in a higher thrust force and higher flux linkage.

The stator of the proposed machine is modular, and the modules are spaced at a certain distance, so better placement of the module becomes very important. It not only increases the coil flux linkage but is also used to minimize the usage of iron and the cost of the machine. GA single variable optimization is used to select the optimal spacing distance between the modules. Complete details of initial values, final values, and the constraint for GA are given in Table 4.

Table 4. Performance of the machine after Genetic Optimization.

Parameters	Initial Value	Final Value	Constraints	TF
S_{TW}	12.24 mm	8.338 mm	7 mm < S_{TW} < 13 mm	136.08 N
S_S	11 mm	13.01 mm	10 mm < S_S < 15 mm	140.601 N

The crooked angle of the stator tooth is analyzed, and its effect on the machine's performance is evaluated. It is the angle made by the inner side of the tooth with the yoke of the stator. Genetic optimization is employed, and it was witnessed that initially when the angle α is increased from 0° to 6.22°, the performance of the machine increases considerably; the best performance being noted at 6.22°. For an angle between 6.22° to 20°, a minute decrease is experienced in the thrust force. Figure 6 shows the crooked angle of the stator tooth, while Figure 7 shows the trend of how thrust force varies with the angle variation.

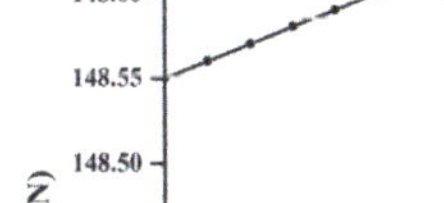

Figure 6. Crooked angle.

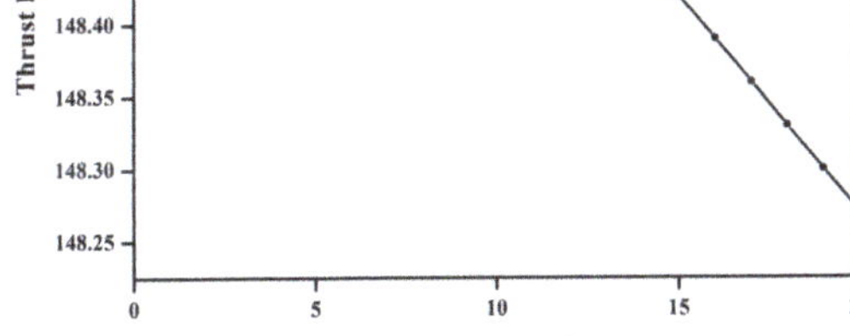

Figure 7. Variation of thrust force with crooked angle.

Figure 8 shows the average thrust force at different armature current starting angles. Three different J_e are considered, and the resultant thrust force is analyzed. It can be seen that the machine performs best at a 0° angle and the thrust force keeps on decreasing, going away from the origin in either direction. GA is used to optimize the starting angle, and its effect on the average thrust force is analyzed. The comparison of before and after optimization flux linkage and thrust force is shown in Figures 9 and 10, respectively.

Table 5 shows before and after optimization values of parameters such as detent force, thrust force, and THD.

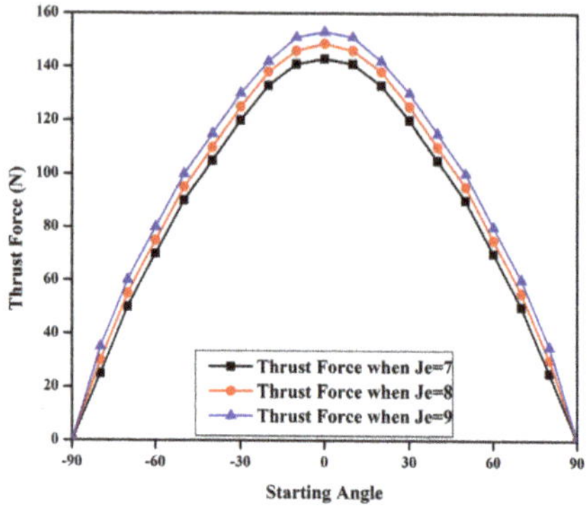

Figure 8. Thrust force at different starting angle.

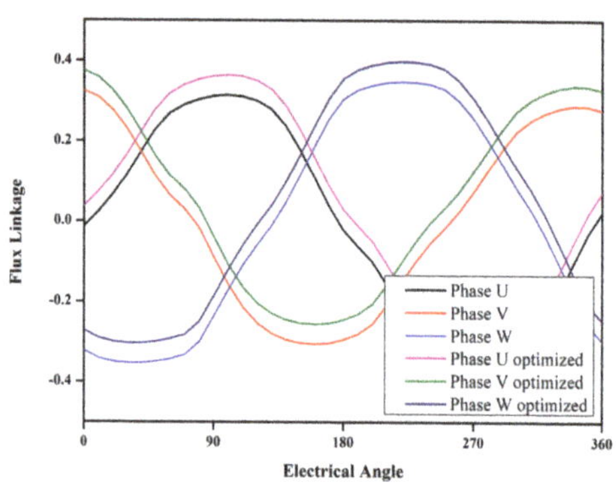

Figure 9. Flux linkage before and after optimization.

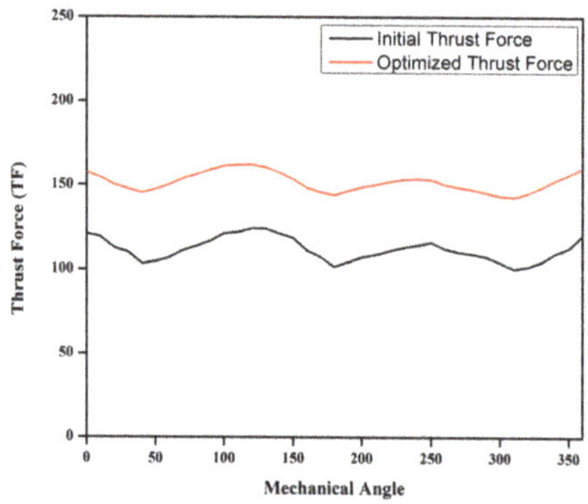

Figure 10. Thrust force comparison before and after optimization.

Table 5. Overall performance of the machine.

Items	Initial Value	After Optimization	Thrust Force Before	Thrust Force After	THD	$F_{(detentp-p)}$
$S.R$	0.225	0.246	129.5353 N	133.4818 N	14.1%	0.2601 N
A_w	2.3 mm	2.6 mm	133.4818 N	135.9811 N	13.7%	0.498 N
S_{TW}	12.24 mm	8.33 mm	135.9811 N	136.08 N	11.79%	2.1 N
S_S	11 mm	13.01 mm	136.08 N	140.60 N	13.21%	4.12 N
α	0	6.22	140.60 N	148.608 N	12.01%	7.2 N
θ	10	0	148.608 N	150.1823 N	12.54%	8.4 N

4. Analysis of Electromagnetic Performance

Parameters such as no-load flux linkage, detent force, total harmonic distortion (THD) of U phase, and thrust force are investigated for a boundary period of 1. No-load flux linkage $\varphi_{(p-p)}$, detent force ($Fd_{(p-p)}$), and thrust force are directly calculated from the FEA. Mathematical calculations were performed to find TF_d for on-load and THDs of no-load flux linkage. Fourier transform of no-load flux linkage is taken, after which Equation (7) is used to find THDs.

$$THD = \frac{\sqrt{\Sigma_{k=2}^{k}\varphi_k^2}}{\varphi_1} \tag{7}$$

where φ_1 represents fundamental component, and φ_2 to φ_k are the harmonics. Thrust force density with respect to mover volume of on-load study is calculated by Equation (8).

$$TF_d = \frac{TF_{average}}{V_m} \tag{8}$$

Figure 11 represent the three-phase no-load flux linkage, all the phases are purely sinusoidal. A difference can be noted between the positive maximum value and the negative value, which points towards the presence of a leakage flux. The flux regulation capability at various DC excitation current is shown in Figure 12. The figure shows that the flux of the proposed machine can be easily controlled by varying the DC current.

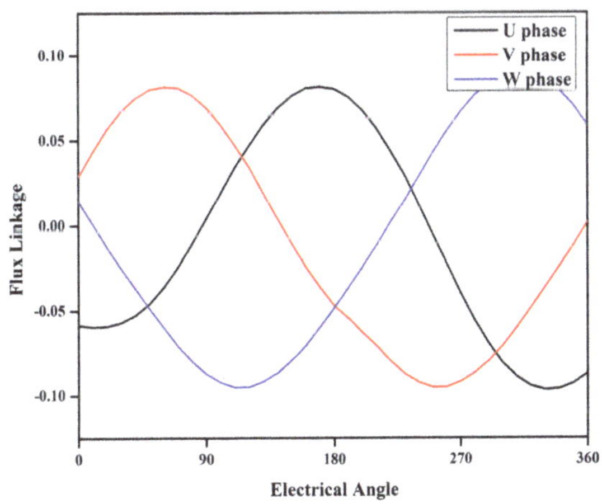

Figure 11. No-load flux linkage.

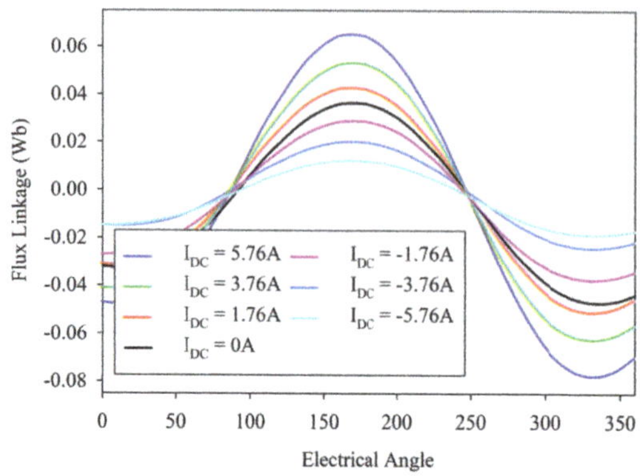

Figure 12. Flux regulation capability of the proposed machine.

Detent force is analyzed in the no-load study when the AC circuit is open, and only DC is fed to the machine. The presence of a magnet makes it the combination of both DC and PM. Depending on its value, it either pulls the machine backward or pushes it forward. Positive detent force helps push the machine forward, while negative detent force pulls the machine backward. This push and pull are the main reasons for thrust force ripple [24,25]. A bipolar detent force can be seen in Figure 13.

The thrust force of the machine is unipolar in nature, as shown in Figure 10. The effect of detent force pull and push can be observed from the thrust force graph. When the value of detent force decreases, i.e., at angle 30° to 120°, 160° to 240°, and 300° to 360°, the thrust force decreases, but when the detent force increases, the thrust force also increases. Table 6 shows the detailed values for THD, peak-to-peak no-load flux linkage, thrust force, and thrust force density.

Performance of the machine is evaluated at different J_a; it can be observed that the thrust force increases linearly up to some extent, but then the linearity is disturbed because the machine is moving toward saturation.

Figure 14 shows thrust force at different J_a and J_e values. Two different values for J_e are considered, and the respective thrust force is shown. Thrust force profiles and power of the machine are evaluated at different velocities and are presented in Figures 15 and 16.

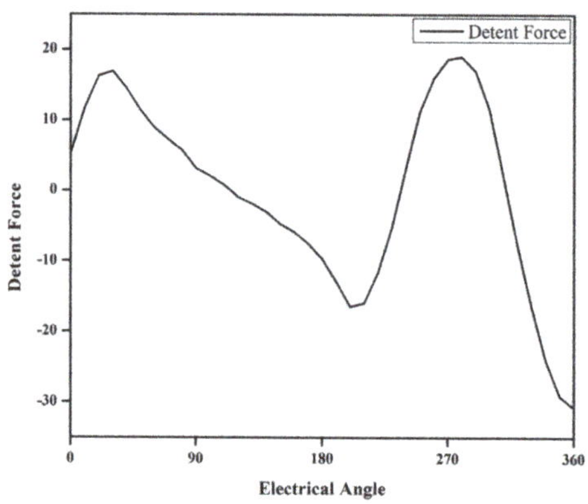

Figure 13. Detent force.

Table 6. Performance indicators.

Parameters	Values
$\varphi_{(p-p)}$	0.12 T
$\varphi_{(+max)}$	0.05 T
$\varphi_{(-max)}$	0.07 T
THD (%)	12.10
$TF_{average}$	129.53 N
$F_{(detentp-p)}$	8.4 N
TF_d	263.64 kN/m^3

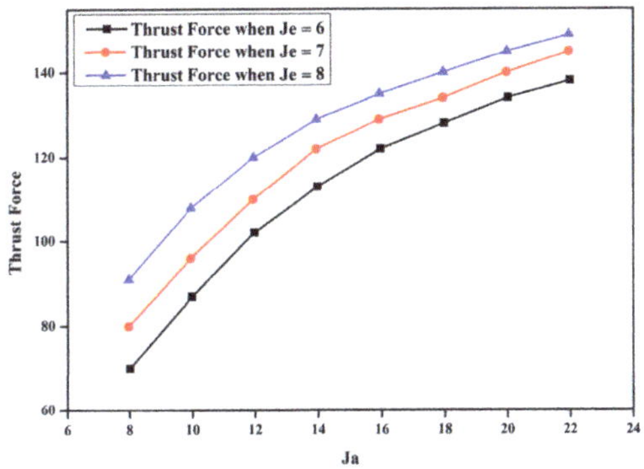

Figure 14. Thrust force at different J_a.

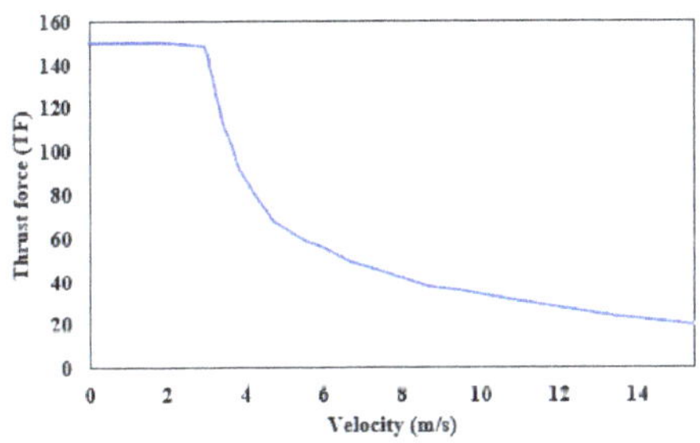

Figure 15. Thrust force at different velocities.

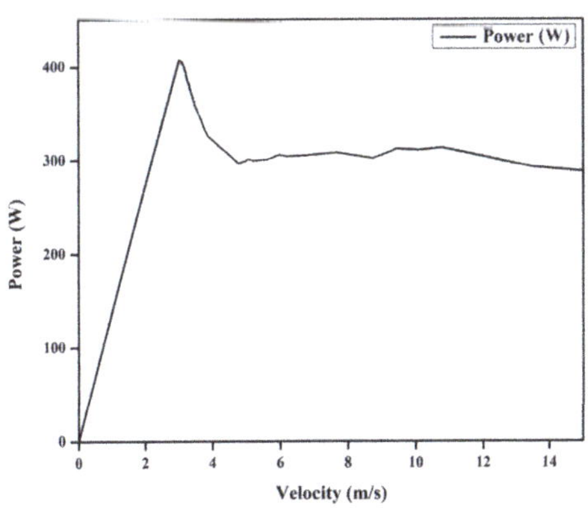

Figure 16. Output power at different velocities.

Different points are considered in the thrust force and velocity graph, and both iron core losses and copper losses are calculated. Overall efficiency of the machine is calculated considering both iron and copper loss. Copper loss of the machine can be calculated using Equation (9).

$$P_c = I\rho JLNQ(1000) \tag{9}$$

where I represents armature current, ρ represents resistivity of the conductor, L is the length of the wire, J_a is the current density of the wire, N is the number of conductors, and Q denotes number of slot pairs.

Iron losses of the machine are calculated directly from JMAG simulation, and the overall efficiency of the machine is calculated. Figure 17 shows losses and overall efficiency of the machine at different velocities. Use of a modular stator reduces the iron volume, thus lowers the iron core losses, increasing the efficiency of the machine. A total of 13% improvement in efficiency is noted using a modular stator because of the iron losses minimization.

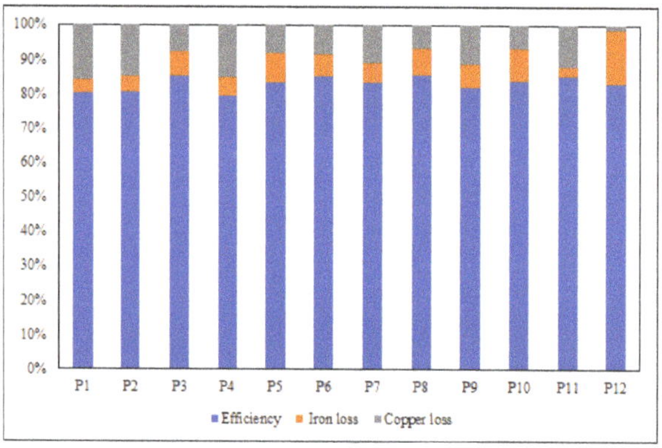

Figure 17. Efficiency at different points.

5. Thermal Analysis

Thermal analysis of the machine has three broad steps: the first one is calculation of three dimensional magnetic losses, thermal analysis, and comparing analytically with lumped parametric model equivalent circuit (LPMEC). The whole process of thermal analysis is shown in the form of a flowchart in Figure 18. The whole procedure of analysis and model building is defined by [20].

5.1. Magnetic Analysis

Since accuracy of the machine is higher in the 3D model, 3D magnetic analysis is considered. Losses calculated are calculated after necessary condition and material setting, and then the losses are used in the thermal analysis.

5.2. Thermal Analysis

Since thermal properties of the machine greatly depend on the nature of the material, conductivity of each material is calculated along with the specific heat capacity of the materials. Various properties of materials are tabulated in Table 7. Room temperature is considered as a reference for the whole process. Three different types of boundaries were defined for various contact types. For direct contact between the two parts, a contact thermal resistance boundary is defined, and heat transfer boundaries are defined for points where heat transfer in the form of convection takes place. Radiation type of heat transfer is ignored in this study. Figure 19 represents the distribution of temperature throughout the machine design. Most of the heat accumulation takes place in the mover as it contains all the active parts (coils and magnets). The stator remains at room temperature as there is no direct contact between the mover and the stator.

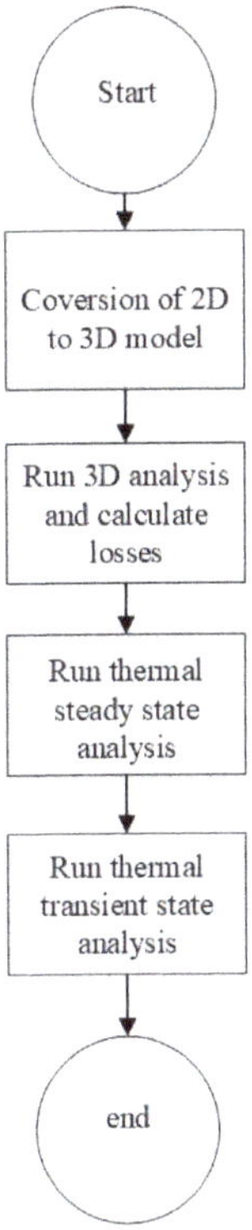

Figure 18. Flowchart of thermal analysis.

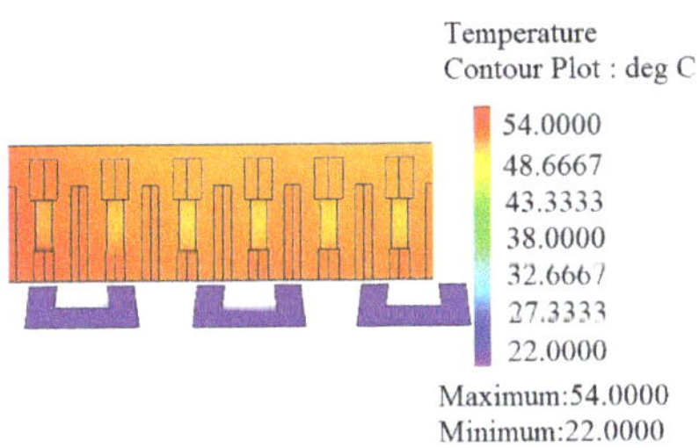

Figure 19. Temperature distribution.

Table 7. Material Properties for thermal analysis.

Part of the Machine	Density (Kg/m^3)	Specific Heat (J/Kg°C)	Thermal Conductivity (W/(m°C))
Modular Stator	7650	460	23
Mover	7650	460	23
Winding coils	4000	380	380

5.3. LPMEC Model

Software simulation of the thermal analysis is validated by an LPMEC model, taking advantage of the heat and electrical system analogy, in which each part of the machine is represented in the form of a resistor, and the value of the resistor depends on its areas and specific heat capacity. A few assumptions were considered while developing this model, and they are given below.

- Only two-dimensional heat flow is considered.
- All the material used in the machine are considered isotropic; heat travels equally in all directions.
- Radiation of the heat is ignored.

The machine is divided into three major parts and each part is then represented by a specific value resistance. Figure 20 represents the various parts of the machine. The coils and magnets are divided into two exactly equal parts while the iron core is represented as single part. Thermal conductivity of each part is presented in Table 8, while resistance of each part can be calculated using Equations (10)–(14).

$$Ri1x = \frac{lix1}{\lambda_{fe} liy1 L_a}, Ri1y = \frac{0.5 * liy1}{\lambda_{fe} lix1 L_a} \tag{10}$$

$$Ri2x = \frac{lix2}{\lambda_{fe} liy2 L_a}, Ri2y = \frac{0.5 * liy2}{\lambda_{fe} lix2 L_a} \tag{11}$$

$$Ri3x = \frac{lix3}{\lambda_{fe} liy3 L_a}, Ri3y = \frac{0.5 * liy3}{\lambda_{fe} lix3 L_a} \tag{12}$$

$$Ri4x = \frac{lix4}{\lambda_{fe} liy4 L_a}, Ri4y = \frac{0.5 * liy4}{\lambda_{fe} lix4 L_a} \tag{13}$$

$$Rm = \frac{lmy}{3\lambda_{PM} lm_x L_a} \tag{14}$$

Equivalent thermal resistance is considered in the case of the slot as the temperature distribution is not uniform because of the insulation and copper winding air gaps. Equation (15) is used to find equivalent thermal resistance.

$$Rcx = \frac{lc_x}{\lambda_{cu} lc_y La}, Rcy = \frac{0.5 lc_y}{\lambda_{cu} lc_x La} \tag{15}$$

Thermal capacity of the machine is represented in the form of capacitance in the LPMEC model, with its values calculated from Equations (16) and (17).

$$C = m \times c_p \tag{16}$$

$$m = \rho \times V \tag{17}$$

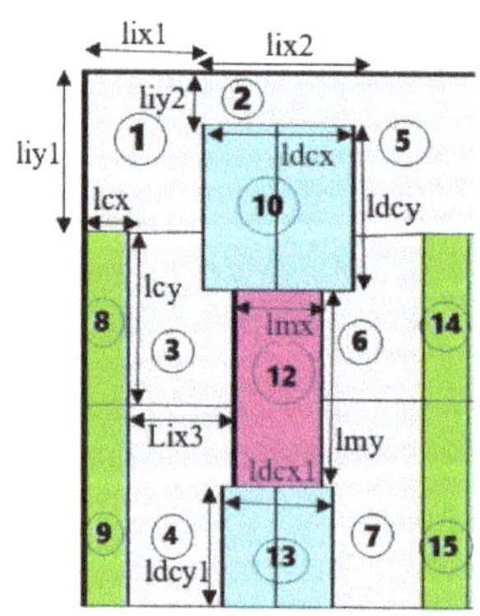

Figure 20. Division of the machine.

Table 8. Thermal conductivity of various parts.

Variable	Thermal Conductivity
λ_{fe}	9 W/m°C
λ_{cu}	28 W/m°C
λ_{PM}	0.68 W/m°C
λ_{cu}	28 W/m°C
Density of iron ρ_i	7870 Kg/m^3
Density of copper ρ_c	8940 Kg/m^3
Density of ferrite magnet ρ_{PM}	5000 Kg/m^3

The developed LPMEC model presented in Figure 21 is then simulated through MATLAB Simulink software, and temperature rise of each part and the temperature rise of the whole machine are observed, shown in Figure 22.

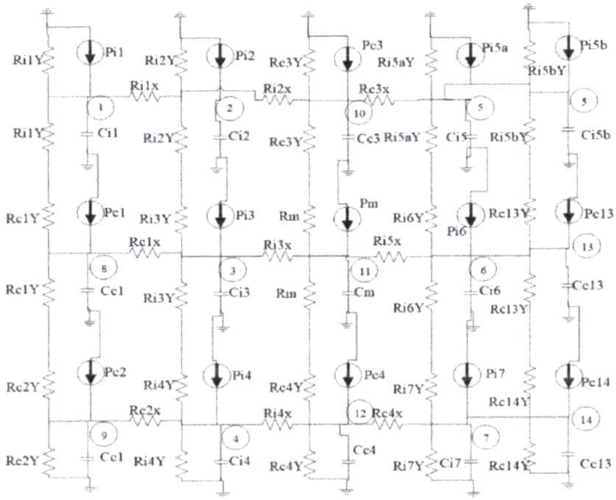

Figure 21. LPMEC model.

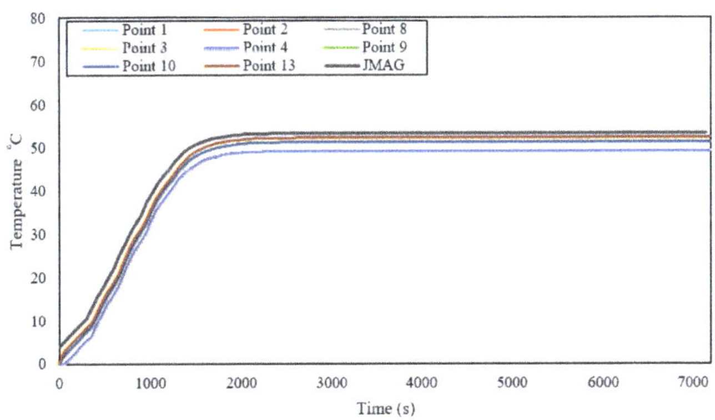

Figure 22. Thermal analysis of each part.

5.4. Comparison of FEA and LPMEC

Analysis of both FEA and LPMEC is compared, and a very good agreement between the two is observed. Figure 22 shows thermal analysis of each part, according to the division mentioned earlier, observed and compared with the average temperature rise of FEA analysis. Figure 23 represents the comparison between the average temperature rise from both methods. Overall, a relative difference of less than 4% is observed, which is quite negligible, since the FEA analysis is performed in three dimensions and is therefore more accurate, while in the case of LPMEC, we have only a two-dimensional flow of temperature.

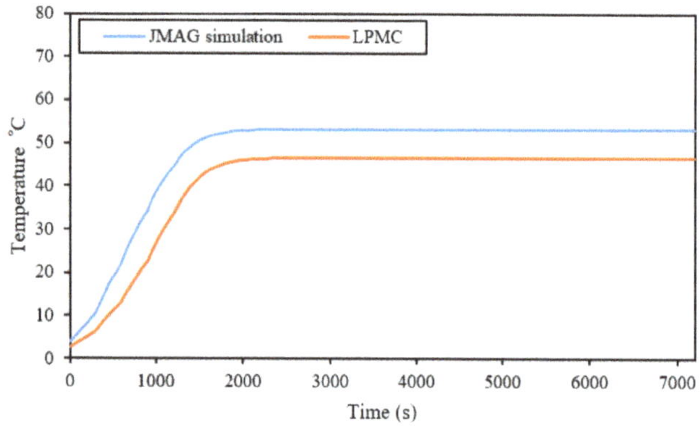

Figure 23. Average temperature rise comparison.

6. Comparison with Conventional LHFSPM

Finally, the proposed LHFSPM is compared to the conventional design [26] in Table 9, which includes a complete comparison. Both the machines use the same dimensions and same type of PM, while the proposed machine achieves the same performance by reducing the volume of PM by 25.56%.

Table 9. Comparison of proposed and conventional LHFSPM.

Parameter	Proposed	[26]
Mover length	131 mm	
Stack length	90 mm	
Airgap	0.8 mm	
Rated velocity	4 m/s	
DC current	5.76 A	6 A
AC turns	230	276
DC turns	81	81
PM type	Ferrite	
PM volume	41.24 cm^3	55.4 cm^3
Thrust force	150.18 N	149.45 N

7. Conclusions

In this paper, a LHFSM with a crooked tooth modular stator is proposed. Electromagnetic performance parameters such as no-load flux linkage, detent force, TF, and thrust force profile at different velocities were studied and analyzed. The crooked tooth technique was devised to improve the thrust force of the machine, and its effect at different angles was presented. Geometric optimization technique was used to enhance the thrust force and mitigate detent force and THD of the proposed machine. GA technique was used to optimize the stator tooth, starting angle, and crooked angle. TF of the machine was improved from an initial value of 129.53 N to an optimized value of 150.1823 N, TF_d was increased from 243.3 kN/m^3 to 306.61 kN/m^3, $\varphi_{(p-p)}$ was increased from 0.12 T to 0.14 T, $Fd_{(p-p)}$ was increased from 0.2601 N to 8.4 N, and the value for THD was slightly reduced from the initial value. Iron losses for the machine were minimized significantly by the use of the modular stator and optimal placement of the modules. In the end, thermal analysis of the machine was performed and was then validated by the LPMEC model. A relative percentage error of 3% was observed.

Author Contributions: Conceptualization, H.U.J. and F.K.; methodology, H.U.J.; software, H.U.J., B.U. and M.Q.; validation, H.U.J., B.U. and M.Q.; resources, F.K., A.H.M. and A.A.A.; original draft preparation, H.U.J.; review and editing, B.U. and F.K.; supervision, F.K.; project administration, F.K. and A.H.M.; funding acquisition, A.H.M. and A.A.A. All authors have read and agreed to the published version of the manuscript.

Funding: There is no external funding.

Data Availability Statement: Not applicable.

Acknowledgments: This work was supported by COMSATS University Islamabad, Abbottabad Campus.

Conflicts of Interest: The authors declare no conflicts of interest.

Abbreviations

AC_w	AC winding slot area
DC_w	DC winding slot area
DSLFSM	Double-sided LFSM
$G.A.$	Genetic algorithm
HEDSLFSM	Hybrid excited double-sided linear flux switching machine
HELFSMs	Hybrid excited LFSMs
HEVs	Hybrid electric vehicles
LFSMs	Linear flux switching machines
LIMs	Linear induction machines
LPMC	Lumped parametric magnetic equivalent circuit
PMLFSMs	Permanent magnet LFSMs
$S.R$	Split ratio
SLFSM	Single-sided LFSM
S_{TW}	Stator tooth width
TF	Thrust force

References

1. Hu, D.; Xu, W.; Dian, R.; Liu, Y.; Zhu, J. Loss minimization control of linear induction motor drive for linear metros. *IEEE Trans. Ind. Electron.* **2017**, *65*, 6870–6880. [CrossRef]
2. Lv, G.; Zeng, D.; Zhou, T.; Liu, Z. Investigation of forces and secondary losses in linear induction motor with the solid and laminated back iron secondary for metro. *IEEE Trans. Ind. Electron.* **2016**, *64*, 4382–4390. [CrossRef]
3. Zou, J.; Xu, W.; Yu, X., Liu, Y.; Ye, C. Multistep model predictive control with current and voltage constraints for linear induction machine based urban transportation. *IEEE Trans. Veh. Technol.* **2017**, *66*, 10817–10829. [CrossRef]
4. Lv, G.; Zhou, T.; Zeng, D. Influence of the ladder-slit secondary on reducing the edge effect and transverse forces in the linear induction motor. *IEEE Trans. Ind. Electron.* **2018**, *65*, 7516–7525. [CrossRef]
5. Sokolov, M.; Saarakkala, S.E.; Hosseinzadeh, R.; Hinkkanen, M. A dynamic model for bearingless flux-switching permanent-magnet linear machines. *IEEE Trans. Energy Convers.* **2020**, *35*, 1218–1227. [CrossRef]
6. Cao, R.; Cheng, M.; Mi, C.C.; Hua, W. Influence of leading design parameters on the force performance of a complementary and modular linear flux-switching permanent-magnet motor. *IEEE Trans. Ind. Electron.* **2013**, *61*, 2165–2175. [CrossRef]
7. Huang, W.; Hua, W.; Yin, F.; Yu, F.; Qi, J. Model predictive thrust force control of a linear flux-switching permanent magnet machine with voltage vectors selection and synthesis. *IEEE Trans. Ind. Electron.* **2018**, *66*, 4956–4967. [CrossRef]
8. Huang, X.Z.; Yu, H.C.; Zhou, B.; Li, L.Y.; Gerada, D.; Gerada, C.; Qian, Z.Y. Detent-force minimization of double-sided permanent magnet linear synchronous motor by shifting one of the primary components. *IEEE Trans. Ind. Electron.* **2019**, *67*, 180–191. [CrossRef]
9. Hao, W.; Wang, Y. Thrust force ripple reduction of two c-core linear flux-switching permanent magnet machines of high thrust force capability. *Energies* **2017**, *10*, 1608. [CrossRef]
10. Song, J.; Dong, F.; Zhao, J.; Lu, S.; Dou, S.; Wang, H. Optimal design of permanent magnet linear synchronous motors based on taguchi method. *IET Electr. Power Appl.* **2017**, *11*, 41–48. [CrossRef]
11. Ma, W.; Wang, X.; Wang, Y.; Li, X. A hts-excitation modular fluxswitching linear machine with static seals. *IEEE Access* **2019**, *7*, 32009–32018. [CrossRef]
12. Wang, Z.-Q.; Long, Z.-Q.; Li, X.-L. Fault analysis and tolerant control for high speed pems maglev train end joint structure with disturbance rejection. *J. Electr. Eng. Technol.* **2019**, *14*, 1357–1366. [CrossRef]

13. Mohamed Jaffar, M.Z. Modeling and Analysis of Saturated Inductances and Torque Ripple Components in Interior Permanent Magnet Motors. Ph.D. Thesis, North Carolina State University, Raleigh, NC, USA, 2019.
14. Ullah, B.; Khan, F.; Milyani, A.H. Analysis of a Discrete Stator Hybrid Excited Flux Switching Linear Machine. *IEEE Access* **2022**, *10*, 8140–8150. [CrossRef]
15. Gas, P.; Kurgan, E. Evaluation of thermal damage of hepatic tissue during thermotherapy based on the arrhenius model. In Proceedings of the 2018 Progress in Applied Electrical Engineering (PAEE), Koscielisko, Poland, 18–22 June 2018; pp. 1–4.
16. Zhao, J.; Guan, X.; Li, C.; Mou, Q.; Chen, Z. Comprehensive evaluation of inter-turn short circuit faults in pmsm used for electric vehicles. *IEEE Trans. Intell. Transp. Syst.* **2020**, *22*, 611–621. [CrossRef]
17. Li, G.; Ojeda, J.; Hoang, E.; Gabsi, M.; Lecrivain, M. Thermal–electromagnetic analysis for driving cycles of embedded flux-switching permanent-magnet motors. *IEEE Trans. Veh. Technol.* **2011**, *61*, 140–151. [CrossRef]
18. Thomas, A.; Zhu, Z.; Li, G. Thermal modelling of switched flux permanent magnet machines. In Proceedings of the 2014 International Conference on Electrical Machines (ICEM), Berlin, Germany, 2–5 September 2014; pp. 2212–2217.
19. Cai, X.; Cheng, M.; Zhu, S.; Zhang, J. Thermal modeling of fluxswitching permanent-magnet machines considering anisotropic conductivity and thermal contact resistance. *IEEE Trans. Ind. Electron.* **2016**, *63*, 3355–3365. [CrossRef]
20. Jan, H.U.; Khan, F.; Ullah, B.; Qasim, M.; Khan, M.A.; Hafeez, G.; Albogamy, F.R. Design and Thermal Modeling of Modular Hybrid Excited Double-Sided Linear Flux Switching Machine. *Energies* **2021**, *14*, 8511. [CrossRef]
21. Jan, H.U.; Khan, F.; Qasim, M.; Ullah, B.; Yousaf, M.; ul Islam, Z. Thermal and Stress Analysis of Linear Hybrid Excited Flux Switching Machine with Modular Stator. In Proceedings of the 2021 International Conference on Emerging Power Technologies (ICEPT), Topi, Pakistan, 10–11 April 2021; pp. 1–6. [CrossRef]
22. Ur Rahman, L.; Khan, F.; Khan, M.A.; Ahmad, N.; Khan, H.A.; Shahzad, M.; Ali, S.; Ali, H. Modular rotor single phase field excited flux switching machine with non-overlapped windings. *Energies* **2019**, *12*, 1576. [CrossRef]
23. Liang, J.; Ming, Z.; Ji, X. Feeding apparatus directly driven by optimal topology flux switching permanent magnet linear motor. *Electr. Power Components Syst.* **2019**, *47*, 903–913. [CrossRef]
24. Zeng, Z.; Shen, Y.; Lu, Q.; Wu, B.; Gerada, D.; Gerada, C. Investigation of a partitioned-primary hybrid-excited flux-switching linear machine with dual-pm. *IEEE Trans. Ind. Appl.* **2019**, *55*, 3649–3659. [CrossRef]
25. Zhao, J.; Mou, Q.; Guo, K.; Liu, X.; Li, J.; Guo, Y. Reduction of the detent force in a flux-switching permanent magnet linear motor. *IEEE Trans. Energy Convers.* **2019**, *34*, 1695–1705. [CrossRef]
26. Hwang, C.C.; Li, P.L.; Liu, C.T. Design and analysis of a novel hybrid excited linear flux switching permanent magnet motor. *IEEE Trans. Magn.* **2012**, *48*, 2969–2972. [CrossRef]

energies

MDPI

Article

Performance Analysis and Optimization of a Novel Outer Rotor Field-Excited Flux-Switching Machine with Combined Semi-Closed and Open Slots Stator

Siddique Akbar [1,*], Faisal Khan [1], Wasiq Ullah [1], Basharat Ullah [1], Ahmad H. Milyani [2,3] and Abdullah Ahmed Azhari [4]

1 Department of Electrical and Computer Engineering, COMSATS University Islamabad, Abbottabad 22060, Pakistan
2 Department of Electrical and Computer Engineering, King Abdulaziz University, Jeddah 21589, Saudi Arabia
3 Center of Research Excellence in Renewable Energy and Power Systems, King Abdulaziz University, Jeddah 21589, Saudi Arabia
4 The Applied College, King Abdulaziz University, Jeddah 21589, Saudi Arabia
* Correspondence: engrsiddique97@gmail.com

Abstract: Slotting effect in electric machines reduces flux per pole that effect magnetic flux density distribution in the air gap which induces harmonics in magnetic flux density causing flux pulsation, that in turn generates dominant torque pulsation in the form of cogging torque and torque ripples. To overcome the abovesaid demerits, a novel outer rotor field-excited flux-switching machine (OR-FSFSM) with a combined semi-closed and open slots stator is proposed in this study. The developed OR-FEFSM offers a high-power factor, due to the utilization of the semi-closed slot for armature coils. The open slot stator structure was chosen for the field excitation coil, which effectively suppresses leakage reluctance that causes flux pulsation. Thus, the influence of torque ripples is reduced, and the average torque is improved. In order to investigate the effectiveness of the proposed OR-FEFSM, a detailed study of stator slot and rotor pole combinations are performed. Based on simplified mathematical formulation, 12S/7P (stator slot/rotor poles), 12S/11P, 12S/13P, and 12S/17P are the most feasible combinations. Finite Element Analysis (FEA) based on comprehensive electromagnetic performance is performed on each combination, and found that 12S/13P offers the highest average torque of 4.62 Nm, whereas 3.72 Nm, 2.72Nm, and 1.68 Nm average torque is offered by 12S/17P, 12S/7P, and 12S/11P, respectively. Based on the initial analysis, 12S/13P was considered for further analysis and optimized using JMAG built-in Genetic Algorithm (GA). Moreover, thermal analysis was performed, and the proposed design was compared with the conventional design.

Keywords: field-excited machine; flux-switching machine; finite element analysis; genetic algorithm; thermal analysis; performance analysis; optimization

Citation: Akbar, S.; Khan, F.; Ullah, W.; Ullah, B.; Milyani, A.H.; Azhari, A.A. Performance Analysis and Optimization of a Novel Outer Rotor Field-Excited Flux-Switching Machine with Combined Semi-Closed and Open Slots Stator. *Energies* **2022**, *15*, 7531. https://doi.org/10.3390/en15207531

Academic Editors: Vladimir Prakht, Mohamed N. Ibrahim and Vadim Kazakbaev

Received: 16 September 2022
Accepted: 9 October 2022
Published: 12 October 2022

1. Introduction

Flux-switching machines (FSMs) are a class of synchronous machines that attract researcher interest due to their doubly salient structure, featuring both excitation sources (permanent magnet (PM) or winding) on the stator and robust rotor structure [1–5]. Since there are no PMs or windings on the rotor, FSM has a robust rotor structure like switched reluctance machines (SRM), making it a strong candidate in many industrial applications where high speed and rigidness are required [6–8]. Thus, due to the inherent feature of high torque and power density, single-phase, three-phase, and multiphase FSM topologies, such as PM excited, field excited (FE), and hybrid excited (HE), have been researched over the past few decades [9,10].

Permanent magnet flux-switching machines (PMFSMs) offer high torque density, efficiency, lower harmonic components in their back-electromotive force (back-EMF), and robust rotor structure [11–13]. Due to the stator's flux-concentrating effect, PMFSM easily generates a high magnetic flux density in the air gap and therefore achieves high torque [14–16]. However, field weakening is a problem for these machines because their d-axis inductance is lowered due to deep magnetic saturation in the stator, limiting their constant power speed range. Additionally, the possibility of irreversible demagnetization, cost of rare earth magnets, uncontrollable flux, and PM's low mechanical strength are all critical issues limiting PM machines to some extent. Therefore, reduced PM or no PM machines are preferred [17,18].

Hybrid excited flux-switching machines (HEFSMs) have the potential to improve power density, efficiency, and flux weakening performance [19–21], and overall reduce the volume of PM by adding FE coils (FEC). However, the presence of both excitation sources (PM and FEC) on the stator makes its geometry complex. To overcome the aforesaid demerits of HEFSM and PMFSM, field-excited flux-switching machines (FEFSMs) are introduced that are fully excited by windings and have more robust rotors than other types of FSM. Despite their lower efficiency, the field current regulation makes FEFSM a suitable alternative for variable speed applications [4,22].

FEFSM that utilize only field winding as a magnetic excitation are proposed in [23]. The field-weakening capabilities of this machine have been greatly enhanced due to flux regulation capability with dc excitation current that effectively reduces excitation magnetic field. However, the field windings are wound around every other stator tooth, causing the magnetic field to be radially excited and unable to be concentrated, which results in low torque density. To improve torque-power density, the authors of [23] investigated wound field-excited FSM with non-overlapped winding and segmented rotor structure. Due to the advantages of flexible flux weakening capability and short flux path, high-torque density is achieved. However, FEFSM topologies with segmented rotors suffer from mechanical constraints, and are not feasible for high-speed applications due to rotor segmentation.

The authors of [24] introduce a partitioned stator FEFSM (PS-FEFSM) with a double stator, having field and armature winding on a separate stator, so that FEC are arranged in the inner stator. In contrast, armature windings are placed in outer stators, thus increasing the torque density by 19% more than conventional FEFSM. However, manufacturing cost increases, due to the partitioned stator with two air gaps; and with metallic coupling of the rotor pieces, the cage losses increase, which can be incorporated with non-metallic sticks.

Despite the average torque improvements, the influence of the cogging torque and torque ripples are dominant in FEFSM. In this regard, the authors of [23] comparatively investigated FEFSM with and without skew rotor for torque ripple reduction. However, torque ripple is suppressed at the cost of a reduction in average torque. To further improve the torque profile, the authors of [25] examined torque characteristics of an OR-FEFSM with maximum torque control; however, this design suffers from magnetic saturation under higher currents. Moreover, dual-stator OR-FEFSM for unbalanced shaft magnetic force in [26] is effectively suppressed at 3% reduction in average torque.

To address the aforementioned problems in various state-of-the-art designs, this paper proposes a novel OR-FEFSM (as shown in Figure 1) for direct-drive embedded in-wheel systems, with an improved torque profile utilizing combined semi-closed and open stators. Furthermore, due to the elimination of mechanical gearboxes, the proposed OR-FEFSM omits maintenance costs, ultimately improving system operation efficiency. In contrast, combined semi-closed and open slots dominantly suppress torque ripple issues and improve overload capability. Therefore, this paper intends a finite element analysis (FEA)-based detailed investigation of the proposed OR-FESM with possible rotor pole combinations.

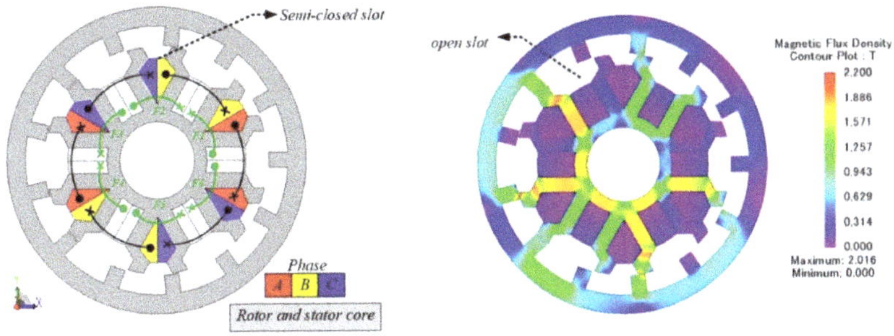

(a) (b)

Figure 1. Proposed OR-FEFSM (**a**) cross-sectional view and (**b**) magnetic flux density map.

Section 2 discusses design topology and operating principles, while Section 3 investigates FEA-based performance analysis of possible rotor pole combinations. Section 4 describes the genetic algorithm. Section 5 examines the best rotor pole's detailed FEA-based performance study. Section 6 depicts a quantitative comparison of electromagnetic performance, whereas Section 7 investigates a comparison of conventional and proposed designs. Section 8 draws a conclusion.

2. Design Topology and Operating Principles

2.1. Design Topology

The proposed 12S/13P OR-FEFSM is shown in Figure 1. It consists of a salient outer rotor and an inner stator having both armature and field winding. The material used for the stator and rotor is 35H210 silicon steel, and copper is used for the winding. Geometric parameters of the proposed OR-FEFSM are shown in Figure 2 and listed in Table 1, while the simplified operating principle based on the flux-switching principle and possible rotor pole combinations are explored as follows.

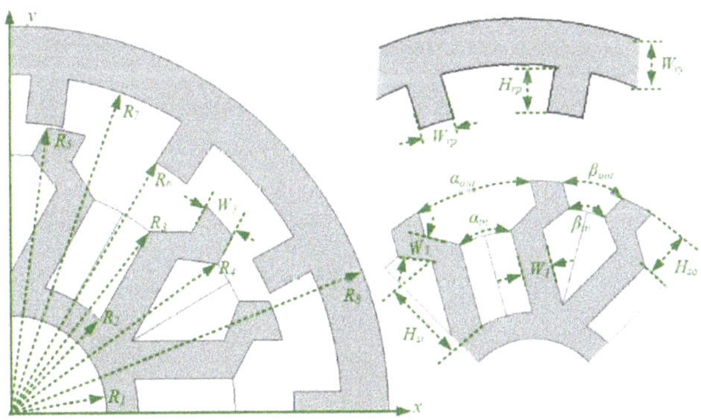

Figure 2. Depicted design parameters of proposed OR-FEFSM.

2.2. Operating Principles

The operating principle of OR-FEFSM is similar to PMFSM, and can be easily explained by the structure shown in Figure 3a,b. The upper laminated part is a rotor-like SRM, and the lower part consists of an armature winding (AW) and FEC labeled in the stator. The flux generated by the FEC flows through the path with the least reluctance. As shown in Figure 3a, when the rotor poles align with the stator teeth, around which the phase coil is wound, the flux that is linked in the coil leaves the rotor tooth and enters the coil. Similarly, as the rotor rotates further and another rotor pole aligns with the next stator tooth of the same coil, flux leaves the coil and enters the rotor tooth, maintaining the same amount of flux-linkage but reversing the polarity, as shown in Figure 3b. The coils flux linkage varies regularly as the rotor rotates, producing sinusoidal back-EMF. Torque is generated when current is appropriately provided to the coils, which drives the armature forward.

Table 1. Design parameter of proposed OR-FEFSM.

Symbol	Value (unit)	Symbol	Value (unit)
R_1	15 mm	W_{ry}	7.5 mm
R_2	20 mm	W_{rp}	6 mm
R_3	35.5 mm	H_{rp}	7.5 mm
R_4	40 mm	r_{out}	22 mm
R_5	44.5 mm	α_{in}	10 mm
R_6	45 mm	W_{sp}	6 mm
R_7	52.5 mm	H_{si}	15.5 mm
R_8	60 mm	H_{so}	9 mm
β_{in}	6.76 mm	β_{out}	11.67 mm
W_1	6 mm	W_2	6 mm
W_3	6 mm	Stack length	45 mm

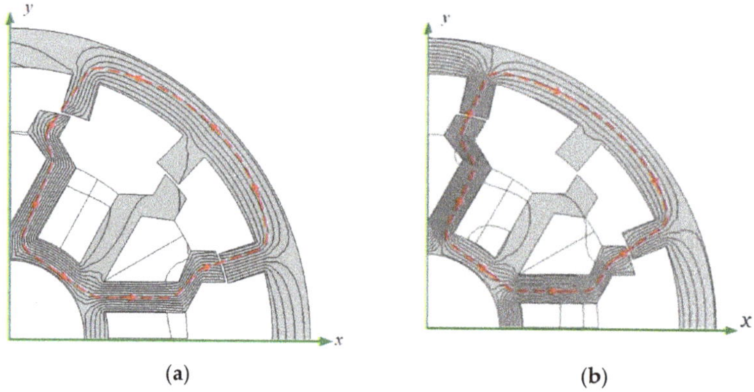

Figure 3. Operating principle of OR-FEFSM (**a**) positive maximum flux and (**b**) negative maximum flux.

2.3. Rotor Poles Combination

Rotor pole position determines the basic principle of FEFSM due to magnetic flux linkage in armature winding that will be positive and negative, depending on the rotor position. For inner and outer rotor FEFSM, possible rotor poles combinations are computed as:

$$N_r = N_s \left(\frac{1 \pm k}{2q} \right) \quad (1)$$

where N_r represents the number of rotor poles, N_s stator slot, K is natural number, and q is number of pairs of phases. For the proposed OR-FEFSM, the values of q and N_s are 6 and 12, respectively, whereas feasible stator slot and rotor pole obtained are 12S/7P, 12S/11P, 12S/13P, and 12S/17P.

Furthermore, for all possible rotor poles combinations, the number of turns per phase (N_a) and number of turns for field winding (N_e) varies based on the armature current density (J_a) and field excitation current density (J_e) as follows:

$$J_a = \frac{I_a N_a}{\alpha S_a} \quad (2)$$

$$J_e = \frac{I_e N_e}{\alpha S_e} \quad (3)$$

where α represent the filling factor while S_a and S_e represent armature and field slot area, respectively. Thus, to decide an optimal values for the aforesaid associated parameters (J_e & J_a), a detailed electromagnetic performance is investigated.

3. Performance Analysis of Feasible Rotor Pole Combination

In this section, a comparative analysis of the proposed OR-FEFSM is conducted for the possible rotor pole combinations. The best design based on electromagnetic analysis will be proceeded for further study.

3.1. No Load Flux Linkage

Figure 4 shows the no-load flux linkage of all feasible rotor pole combinations. It is worth noting that the phase flux linkage obtained under no-load operation with $J_e = 15\ \text{A/mm}^2$. Analysis reveals that OR-FEFSM with 12S-7P design offers the highest flux linkage of 0.020 Wb, which is 54%, 7.15%, and 37.97% higher than 11, 13, and 17 rotor poles, respectively.

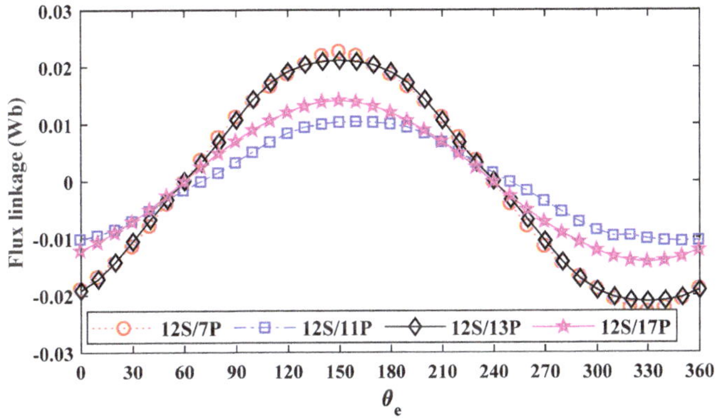

Figure 4. No-load operation flux linkages of all feasible combinations.

3.2. No Load Back-EMF

Under no-load operation, phase back-EMF for 12S/7P, 12S/11P, 12S/13P, and 12S/17P through 2D FEA is shown in Figure 5. Analysis reveals that 12S/13P shows the highest back-EMF of 44.43 V, whereas the phase back-EMF for 12S/7P, 12S/11P, and 12S/17P are 25 V, 18.13 V, and 38.63 V, respectively.

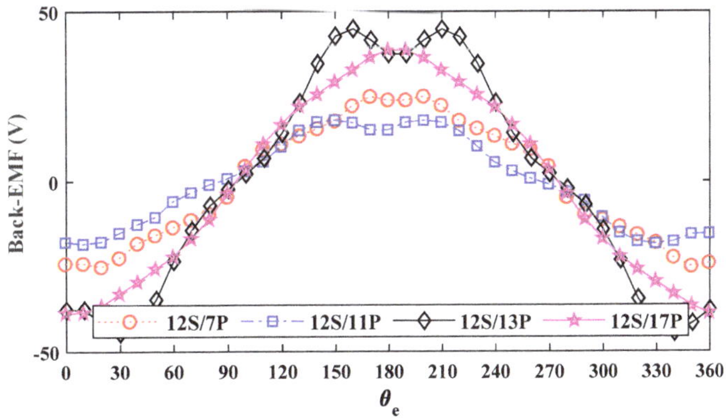

Figure 5. No-load operation phase back-EMF.

3.3. Cogging Torque

The torque obtained is cogging torque when no current is applied to the armature winding. This torque is due to slotting effects that cause acoustic noise and vibration, and is therefore undesirable. Since slotting effects change with the rotor pole number variation, the resultant cogging torque greatly vary. The cogging torques of 12S/7P, 12S/11P, 12S/13P, and 12S/17P are illustrated in Figure 6. In regard to peak magnitude, analysis reveals that 12S/17P design offers lowest cogging torque of 0.0508 Nm, whereas 12S/11P shows highest cogging torque of 0.337 Nm. Furthermore, the cogging torque profiles for 12S/13P and 12S/7P show moderate values.

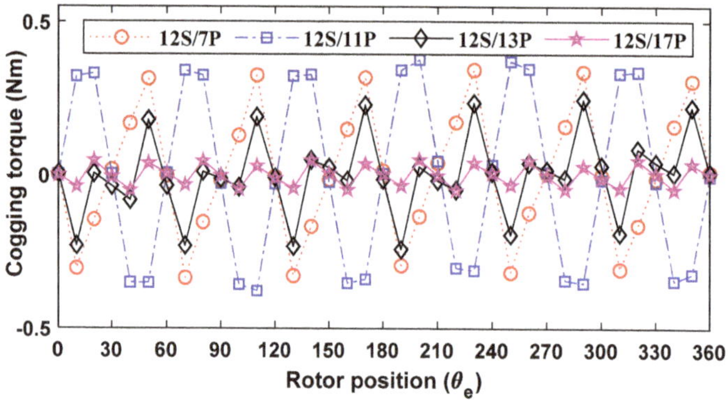

Figure 6. Variation of cogging torque profile.

3.4. Instantenous Torque

Under loaded and rated operating conditions of $J_a = J_e = 15\ \mathrm{A/mm^2}$ and speed of 1500 rpm, instantaneous torque profile is shown in Figure 7. Comparative analysis of various rotor poles illustrates that 12S/13P offers the highest average torque (T_{avg}) of 4.62 Nm, whereas 12S/11P shows the lowest average torque of 1.68 Nm.

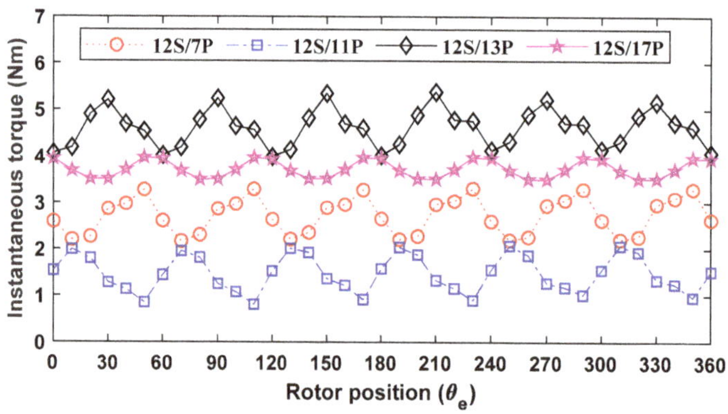

Figure 7. Instantaneous torque profile with rotor pole number.

Table 2 represents average torque, cogging torque, torque ripples ratio, flux linkage, and back-EMF of possible rotor pole combinations. From the above analysis, it is clear that the 12S/13P combination has a more significant torque density and back-EMF at $J_a = J_e = 15\ \mathrm{A/mm^2}$. Therefore, it is considered for further study.

Table 2. Quantitative performance of the OR-FEFSM with various rotor pole numbers at $J_a = J_e = 15\ \text{A/mm}^2$.

	12S/7P	12S/11P	12S/13P	12S/17P
Average Torque (Nm)	2.72	1.68	4.62	3.72
Cogging Torque $_{\text{P-P}}$ (Nm)	0.344	0.377	0.246	0.0508
Torque ripples (%)	32	16	31	12
Flux Linkage (Wb)	0.022816	0.010481	0.021185	0.014152
Back-EMF (V)	25.06	18.13	44.84	38.64

4. Genetic Optimization

In this section, optimization using genetic algorithm (GA) is used for the proposed OR-FEFSM with 12S/13P. GA is a widely common solution for optimization issues because it is particularly good at solving highly non-linear objective functions. The GA-based optimization technique is based on the evolution of the world's animal population. It begins with a random population of variables, similar to a pool of chromosomes. A generation is the term used to describe each iteration of the algorithm. Only a few chromosomes with the most significant fitness values are carried down into the next generation, known as exceptional offspring. In addition, the technique generates new children for the next generation by simulating crossover and mutation, which are binary and unary acts on existing chromosomes, respectively. The method is repeated until one of the end conditions are met [27]. The number of generations and population size are defined to reach an optimum global value of the objective function. Figure 8 illustrates the GA workflow.

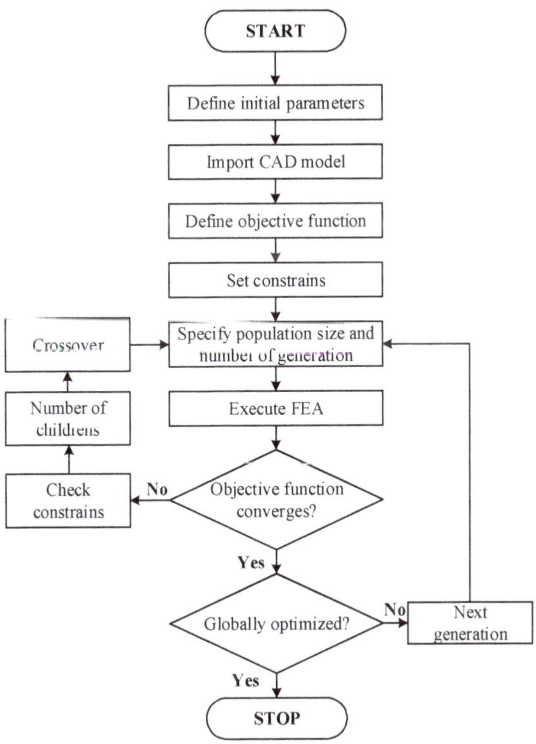

Figure 8. Flow chart of Genetic Optimization.

Geometry editor is used to develop the initial design, and the CAD parameters are then imported into the designer. The objective function and constraints are given below, composed of two sub-objectives. Table 3 describes the ranges and constraints of variable CAD parameters.

$$\text{Objective function} = \begin{cases} \max(T_{avg}) \\ \min(T_{ripples}) \end{cases}$$

$$\text{Constraints} = \begin{cases} T_{avg} > 4.62 \\ T_{ripples} < 0.28 \end{cases}$$

$$\text{Design variables} = \begin{cases} 3 \leq W_{ry} \leq 12 \\ 2 \leq W_{rp} \leq 5 \\ 3 \leq W_1 \leq 8 \\ 3 \leq W_2 \leq 8 \\ 3 \leq W_3 \leq 8 \\ 12 \leq R_1 \leq 15 \end{cases}$$

Table 3. Variables for Genetic Optimization.

Parameters	Initial (mm)	Boundary Conditions	Optimized (mm)
W_{ry}	7.5	$3 \leq W_{ry} \leq 12$	6.3505
W_{rp}	3	$1 \leq W_{rp} \leq 5$	3.3055
W_1	6	$3 \leq W_1 \leq 8$	7.97017
W_2	6	$3 \leq W_2 \leq 8$	5.9480
W_3	6	$3 \leq W_3 \leq 8$	6.8116

JMAG built-in global optimization, which utilizes the GA approach, was used to optimize the geometrically significant parameters. Geometrical design variables, such as the yoke length (W_{ry}), width of rotor tooth (W_{rp}), the area of the armature slot, area of the FEC slot, width of stator poles (W_1, W_2, W_3), and the shaft radius (R_1), were used to determine the optimization problem of the design. During optimization, key dimensions, such as the stator outer radius (R_5), air gap, rotor outer radius (R_8), stack length, rated field and armature current densities, and turns, were kept constant to maintain constant electrical and magnetic loading. In optimization settings, maximum generations were set at 14, population size at 16, number of children at 17, and stopping criteria at 10. The number of elements and nodes of each model were 12,246 and 7672, respectively. The mesh size was set to 0.5. After computing 451 case studies with GA, which took nearly 112 hours, the file size was 110 GB, the PC used was a HP core i5, 2.5 GHz, 8 GB RAM, and optimum values were obtained. Optimization results, such as torque and ripple ratio w.r.t, W_{ry}, W_{rp}, W_1, W_2, W_3, and R_1, are illustrated in Figure 9 in relation to one another. Table 3 depicts the proposed design's global optimal parameters.

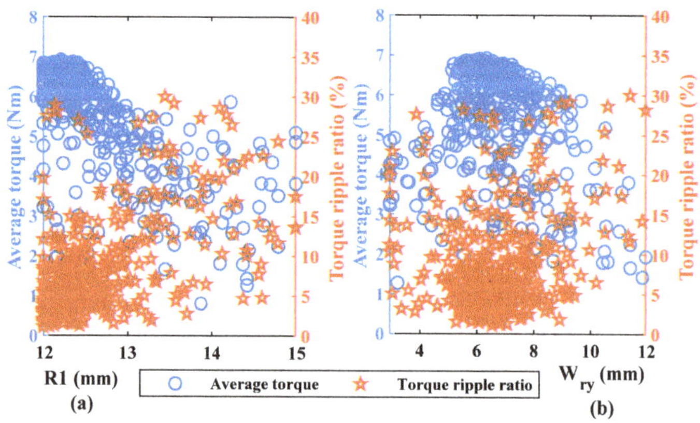

Figure 9. *Cont.*

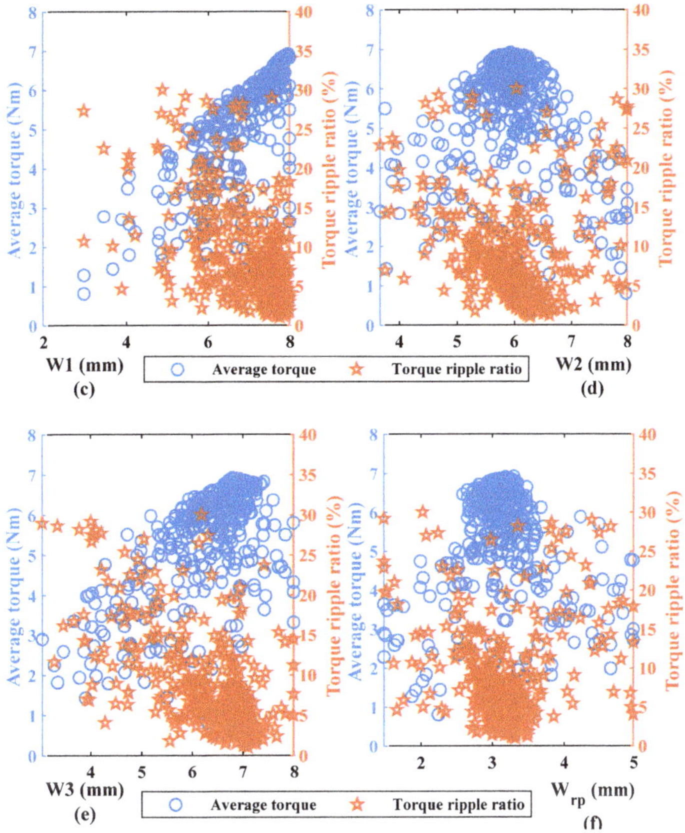

Figure 9. Torque and ripples ratio w.r.t optimization parameters (**a**) R1 (**b**) W_{ry} (**c**) W1 (**d**) W2 (**e**) W3 (**f**) W_r.

5. FEA Based Electromagnetic Performance of Optimized Design

In this section, a detailed comparative analysis of optimal design is discussed.

5.1. No Load Analysis

The no-load airgap flux density of 12S/13P initial and optimized design substantially fluctuates, resulting in a significant variation in the magnetic flux density over one periodic cycle, as shown in Figure 10. The peak value of airgap flux density of optimized design is 30% more than the initial. The variation of the magnetic flux density results in phase flux linkage variation. Figure 11a,b shows no load flux linkage and harmonic spectra of initial and optimized design. Analysis shows that the peak value of flux linkage is improved by 31.15% after optimization. Similarly, odd harmonics and total harmonic distortion (THD) were improved by 41.88 and 58.33%. The formula for calculating THD is shown in (4) below.

$$THD = \frac{\sum_{i=2}^{N} \varnothing_i}{\varnothing_1} \quad (4)$$

where $\varnothing_1$ is fundamental hormonic component and N is natural number

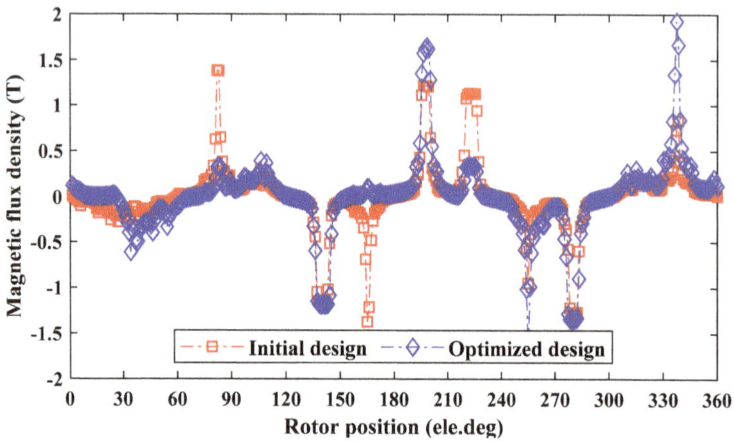

Figure 10. Variation of the magnetic flux density over one periodic cycle.

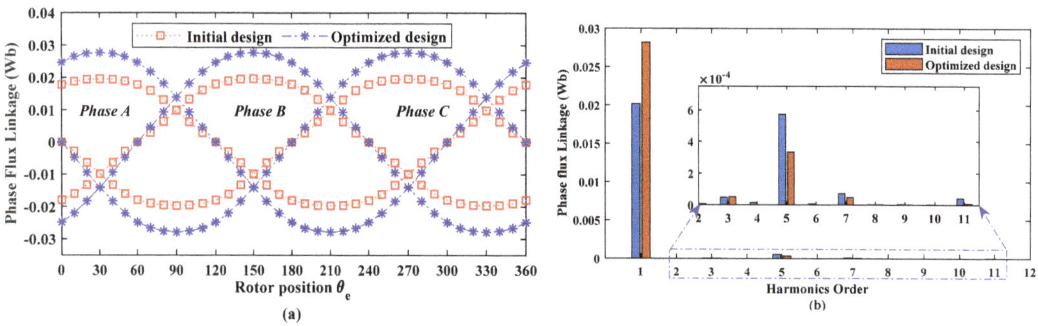

Figure 11. Initial and optimized design phase flux linkage (**a**) waveform and (**b**) harmonic spectra.

Under no-load condition, phase-back-EMF and harmonic spectra for initial and optimized design through 2D FEA is shown in Figure 12a,b. Analysis reveals that back-EMF is improved by 28%, while odd harmonics and THD is improved by 33.2% and 39.63%, respectively, after optimization. Figure 13 shows the comparison between initial and optimized cogging torque; after optimization, the cogging torque is reduced by 54.93%.

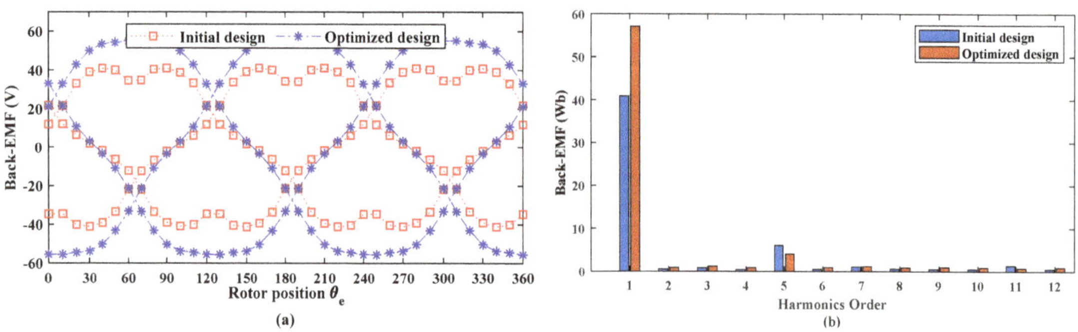

Figure 12. Initial and optimized design back-EMF (**a**) waveform and (**b**) harmonic spectra.

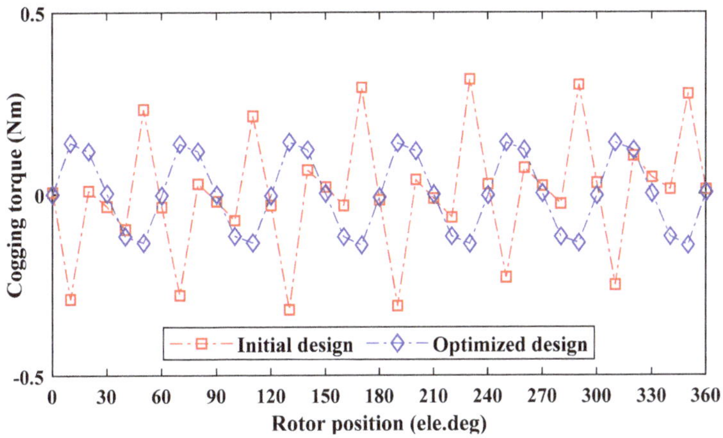

Figure 13. Cogging torque analysis of initial and optimized design.

5.2. Load Analysis

Under loaded and rated operating conditions of $J_a = J_e = 15$ A/mm^2 and speed of 1500 rpm, instantaneous torque profile of 12S/13P is shown in Figure 14, which offers a 41.99% improvement in torque and 87.76% in Torque ripple ratio (T_{rr}). Furthermore, average torque profiles with torque ripples under different electric loading are illustrated in Figure 15, demonstrating the over-load capability of the proposed OR-FEFSM before and after optimization. It can be seen that under all-electric loading conditions, average torque increases, whereas the torque ripple ratio decreases. From the analysis, the torque and ripples ratios are improved by 41.99% and 87.76%, respectively, at the rated conditions. The formula for calculating torque ripple ratio is discussed in (5).

$$T_{rr} = \frac{T_{max} - T_{min}}{T_{avg}} \times 100 \quad (5)$$

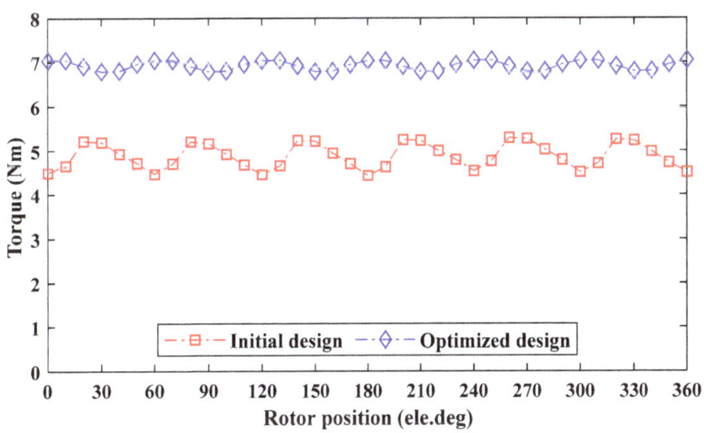

Figure 14. Instantaneous toque profile of initial and optimized design.

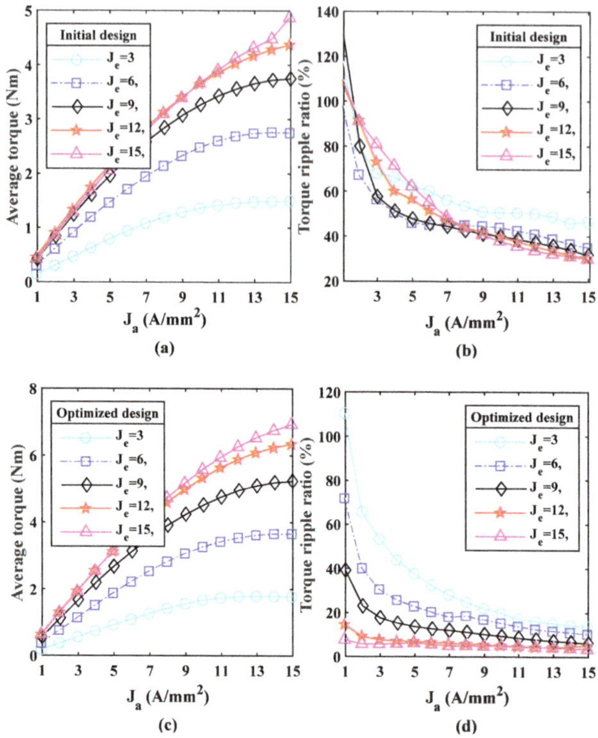

Figure 15. Overload capability of proposed design before and after optimization. (**a**) Torque w.r.t J_a and J_e before optimization; (**b**) ripples w.r.t J_a and J_e before optimization; (**c**) torque w.r.t J_a and J_e after optimization; and (**d**) ripples w.r.t J_a and J_e after optimization.

In (5) T_{max} is the maximum value of torque, T_{min} is the minimum value of torque, and T_{avg} is the average value of torque. Figure 16 shows electromagnetic torque performance of initial and optimized design w.r.t different electrical degrees.

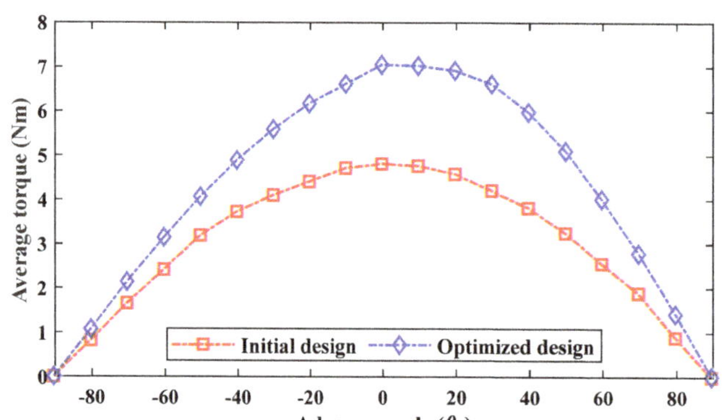

Figure 16. Torque w.r.t different electrical degrees before and after optimization.

5.3. Dynamic Analysis

In designing an electric machine, dynamic characteristics analysis is one of the key studies that provides wide operational capability. In this regard, torque and output power, characteristics curve versus speed and power factor, give a detailed illustration of its behavior under low-speed, high-speed, constant torque, and constant output power regions, which is a prerequisite for direct-drive in-wheel-embedded systems.

The output power of the proposed machine is calculated by multiplying T_{avg} with the corresponding speed. Input power is the summation of output power and total losses, including core and copper losses. Core losses can be calculated from 2D-FEA at the specified points, and (6) is used for copper losses calculation.

$$P_{Copper} = P_{Copper}(AW) + P_{Copper}(FEC) \tag{6}$$

where,

$$P_{Copper\ (AW)} = I_{rms}\rho JL(NQ)(1000) \tag{7}$$

$$P_{Copper\ (FEC)} = I\rho JL(NQ)(1000) \tag{8}$$

In (8), I represent current, ρ is resistivity, J represents current density, N and Q are number of turns and slot pair, respectively.

$$P_{out} = T_{avg} * \omega \tag{9}$$

$$P_{input} = P_{out} + P_{Copper} + P_{Core} \tag{10}$$

Finally, efficiency is given as:

$$\eta = \frac{P_{out}}{P_{input}} 100\ \% \tag{11}$$

The characteristics curve of torque and power versus speed on initial and optimized designs is depicted in Figures 17a and 17b, respectively.

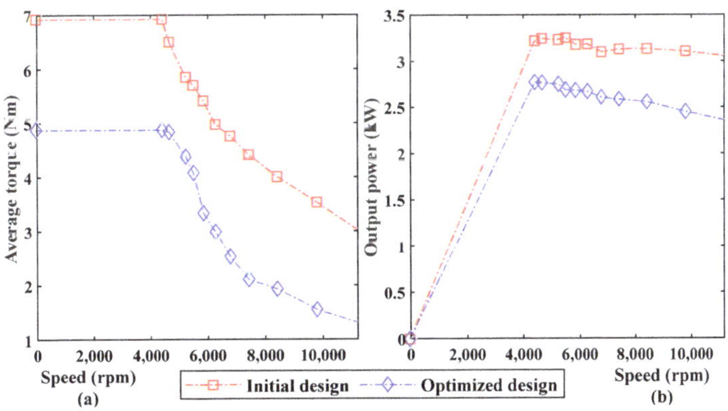

Figure 17. (**a**) Torque vs. speed curve; (**b**) power vs. speed curve.

Similarly, the power factor can be calculated as:

$$P = \frac{3}{2}(V_d I_q + V_q I_d) \tag{12}$$

$$Q = \frac{3}{2}(V_d I_q + V_q I_d) \tag{13}$$

$$S = \sqrt{P^2 + Q^2} \tag{14}$$

$$PF = cos\varnothing = \frac{P}{S} \tag{15}$$

The power factor of initial and optimized design is shown in Figure 18.

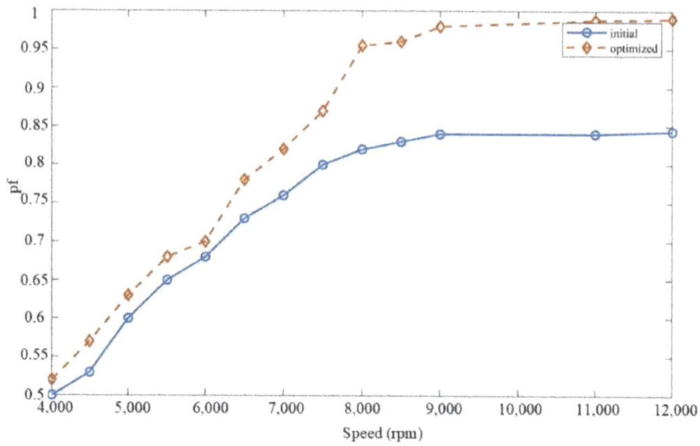

Figure 18. Power factor of initial and optimized design w.r.t different speed.

A comparison of the initial and optimized OR-FEFSM designs is performed in order to verify and examine the performance of the GA optimization technique. Table 4 illustrates a quantitative assessment of the proposed machine.

Table 4. Performance comparison of initial and optimized design.

Parameters	Initial Value	Optimized Value	Improvement
Torque (Nm)	4.877	6.925	41.99%
Torque Ripples (%)	0.30	0.037	87.76%
Flux linkage (Wb)	0.021	0.0276	23.91%
Airgap flux density (T)	1.3886	1.9276	27.94%
$\text{Back} - \text{emf}_{\text{peak}}(\text{V})$	41.07	55.362	25.91%
Cogging torque(Nm)	0.3160	0.1424	54.93%
Odd harmonics in flux (Wb)	0.000573	0.000333	41.88%
THD in flux (%)	2.88	1.2	58.33%
Odd harmonics in back-EMF (V)	6.1206	4.09209	33.2%
THD in back emf (%)	16.4	9.9	39.63%
Output power (Kw)	2.27	3.223	29.56%
Power factor (Pf)	0.84	0.97	13.4%
Efficiency (%)	83.4	87.70	4.9%

6. Thermal Analysis

Thermal analysis is one of the most important analyses in electric machine design, as heat dissipation limits the long-term functionality of the machine. Therefore, it helps to determine the operating limits and the insulating class of the machine. The electromagnetic performance reduces when the temperature exceeds a certain permissible range, resulting in an inter-turn short circuit problem [28]. Current flow through the conductor produces heat dissipation and power loss, which raises the temperature of the whole machine. The machine has core losses in addition to the copper losses associated with the windings, which are due to the eddy current and hysteresis losses in the core. All of these losses behave as a heat source and raise temperatures. Therefore, the magnetic loss study is used to determine the losses, and thermal analysis is used to analyze the temperature distribution.

For the proposed machine, initial losses were determined using a 3D FEA followed by thermal analysis in 3D thermal studies, as 3D analysis is more precise than 2D analysis. To calculate the temperature distribution of the complete machine, the 3D loss study is combined with the 3D heat study. The 3D thermal study reveals that the stator temperature significantly rises due to the presence of all excitation sources on the stator; however, the temperature of the rotor minimally rises, as the proposed machine has a salient rotor. Figure 19 depicts a contour plot of the temperature distribution. In the stator, the temperature distribution in the windings reaches a maximum value of 46 °C. At the rotor, the highest temperature observed is 24 °C degrees.

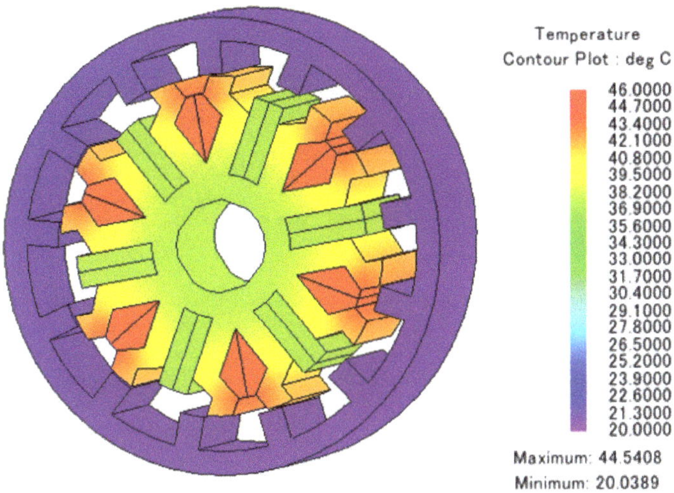

Figure 19. Thermal analysis of the proposed OR-FEFSM.

7. Comparison of Proposed and Conventional Design

In this section, a detailed comparison of the conventional and proposed design is performed. For fair comparison with [23], the proposed model is redesigned and optimized under the same outer dimensions, stack length, current density, and air gap length. The average torque of scaled 12S/13P design is 47.95 Nm which is 36.6%, 57.1%, 48.77%, and 96.52% more than the conventional straight WFSM, skewed WFSM, 12S/8P SRM, and 12S/10P SRM, respectively. Based on the key performance matrix listed in Table 5, without any influence on efficiency, it is clear that the proposed design offers 9.4%, 65.8%, and 63.7% less ripples than the conventional straight WFSM, 12S/8P SRM, and 12S/10P SRM, respectively. Furthermore, individual machine part weight and overall weight of the proposed machine are the lowest, resulting in a better torque density than the aforementioned state-of-the-art model.

Table 5. Performance comparison of conventional and proposed design.

Parameters	Straight WFSM	Skewed WFSM	12/8 SRM	12/10 SRM	Proposed Machine
Average Torque (Nm)	35.1	30.52	32.32	28.4	47.95
Torque Ripple (%)	27.6	6.85	73.1	68.9	25
Efficiency (%)	89.4	88	89.9	90.5	89.3
Weight of iron core (kg)	19.87		19.54	19.74	16.19
Weight of copper (kg)	17.86		13.74	12.63	3.121
Weight in total (kg)	37.73		34.33	32.37	19.311
Torque Density (Nm/kg)	0.930	0.808	0.941	0.877	2.483

8. Conclusions

This paper proposed a novel outer rotor field-excited flux-switching machine with a combined semi-closed and open slots stator to improve power factor and torque profile.

Semi-closed slots utilization improves power factor, whereas open slots suppress slot leakages that effectively suppress leakage reluctance that causes flux pulsation, improving torque ripples, and mean torque. The effectiveness of the proposed OR- FEFSM was investigated through FEA with different stator slot and rotor pole combinations. Initial FEA-based comprehensive electromagnetic performance reveals that 12S/13P offers the highest average torque of 4.62 Nm, which was considered for further study and optimized through GA optimization. The maximum output power obtained is 3.22 kW with an efficiency of 87.7%. Finally, the proposed design was compared with the state-of-the-art model, and the results showed there was significant improvement in average torque, torque ripples, weight of machine, and torque density.

Author Contributions: Conceptualization, S.A., W.U. and F.K.; methodology, S.A. and B.U.; software, S.A. and B.U.; validation, S.A. and W.U.; resources, F.K., A.H.M. and A.A.A.; formal analysis, B.U., A.H.M. and A.A.A.; original draft preparation, S.A. and W.U.; visualization, B.U.; review and editing, B.U. and F.K.; supervision, F.K.; project administration, F.K., A.H.M. and A.A.A.; funding acquisition, A.H.M., A.A.A. All authors have read and agreed to the published version of the manuscript.

Funding: This research received no external funding.

Conflicts of Interest: The authors declare no conflict of interest.

References

1. Ullah, W.; Khan, F.; Sulaiman, E.; Umair, M.; Ullah, N.; Khan, B. Analytical validation of novel consequent pole E-core stator permanent magnet flux switching machine. *IET Electr. Power Appl.* **2020**, *14*, 789–796. [CrossRef]
2. Jiang, W.; Huang, W.; Lin, X.; Zhao, Y.; Zhu, S. Analysis of Rotor Poles and Armature Winding Configurations Combinations of Wound Field Flux Switching Machines. *IEEE Trans. Ind. Electron.* **2021**, *68*, 7838–7849. [CrossRef]
3. Jiang, W.; Huang, W.; Lin, X.; Zhao, Y.; Wu, X.; Zhao, Y.; Dong, D.; Jiang, X. A Novel Stator Wound Field Flux Switching Machine with the Combination of Overlapping Armature Winding and Asymmetric Stator Poles. *IEEE Trans. Ind. Electron.* **2022**, *69*, 2737–2748. [CrossRef]
4. Zulu, A.; Mecrow, B.; Armstrong, M. A Wound-Field Three-Phase Flux-Switching Synchronous Motor with All Excitation Sources on the Stator. *IEEE Trans. Ind. Appl.* **2010**, *46*, 2363–2371. [CrossRef]
5. Tang, Y.; Paulides, J.; Motoasca, T.; Lomonova, E. Flux-Switching Machine with DC Excitation. *IEEE Trans. Magn.* **2012**, *48*, 3583–3586. [CrossRef]
6. Wu, Z.; Zhu, Z.; Wang, C. Reduction of On-Load DC Winding-Induced Voltage in Partitioned Stator Wound Field Switched Flux Machines by Dual Three-Phase Armature Winding. *IEEE Trans. Ind. Electron.* **2022**, *69*, 5409–5420. [CrossRef]
7. Ullah, W.; Khan, F.; Hussain, S. A Comparative Study of Dual Stator with Novel Dual Rotor Permanent Magnet Flux Switching Generator for Counter Rotating Wind Turbine Applications. *IEEE Access* **2022**, *10*, 8243–8261. [CrossRef]
8. Tang, Y.; Motoasca, E.; Paulides, J.; Lomonova, E. Comparison of flux-switching machines and permanent magnet synchronous machines in an in-wheel traction application. *COMPEL-Int. J. Comput. Math. Electr. Electron. Eng.* **2012**, *32*, 153–165. [CrossRef]
9. Othman, S.; Ahmad, M.; Rahim, J.; Bahrim, F.; Sulaiman, E. Design Improvement of Three Phase 12Slot-14Pole Outer Rotor Field Excitation Flux Switching Motor. *Int. J. Power Electron. Drive Syst. (IJPEDS)* **2017**, *8*, 239. [CrossRef]
10. Shen, Y.; Hu, P.; Jin, S.; Wei, Y.; Lan, R.; Zhuang, S.; Zhu, H.; Cheng, S.; Chen, J.; Wang, D.; et al. Design of Novel Shaftless Pump-Jet Propulsor for Multi-Purpose Long-Range and High-Speed Autonomous Underwater Vehicle. *IEEE Trans. Magn.* **2016**, *52*, 1–4. [CrossRef]
11. Su, P.; Hua, W.; Wu, Z.; Chen, Z.; Zhang, G.; Cheng, M. Comprehensive Comparison of Rotor Permanent Magnet and Stator Permanent Magnet Flux-Switching Machines. *IEEE Trans. Ind. Electron.* **2019**, *66*, 5862–5871. [CrossRef]
12. Hua, W.; Cheng, M.; Zhu, Z.; Howe, D. Analysis and Optimization of Back EMF Waveform of a Flux-Switching Permanent Magnet Motor. *IEEE Trans. Energy Convers.* **2008**, *23*, 727–733. [CrossRef]
13. Du, Y.; Zhang, C.; Zhu, X.; Xiao, F.; Sun, Y.; Zuo, Y.; Quan, L. Principle and analysis of doubly salient PM motor with Π-shaped stator iron core segments. *IEEE Trans. Ind. Electron.* **2018**, *66*, 1962–1972. [CrossRef]
14. Ullah, W.; Khan, F.; Hussain, S. A Novel Dual Rotor Permanent Magnet Flux Switching Generator for Counter Rotating Wind Turbine Applications. *IEEE Access* **2022**, *10*, 16456–16467. [CrossRef]
15. Chau, K.; Chan, C.C.; Liu, C. Overview of permanent-magnet brushless drives for electric and hybrid electric vehicles. *IEEE Trans. Ind. Electron.* **2008**, *55*, 2246–2257. [CrossRef]
16. Du, Z.S.; Lipo, T.A. Efficient utilization of rare earth permanent-magnet materials and torque ripple reduction in interior permanent-magnet machines. *IEEE Trans. Ind. Appl.* **2017**, *53*, 3485–3495. [CrossRef]
17. Dorrell, D.; Parsa, L.; Boldea, I. Automotive Electric Motors, Generators, and Actuator Drive Systems with Reduced or No Permanent Magnets and Innovative Design Concepts. *IEEE Trans. Ind. Electron.* **2014**, *61*, 5693–5695. [CrossRef]

18. Boldea, I.; Tutelea, L.N.; Parsa, L.; Dorrell, D. Automotive Electric Propulsion Systems with Reduced or No Permanent Magnets: An Overview. *IEEE Trans. Ind. Electron.* **2014**, *61*, 5696–5711. [CrossRef]
19. Yang, G.; Lin, M.; Li, N.; Tan, G.; Zhang, B. Flux-Weakening Control Combined with Magnetization State Manipulation of Hybrid Permanent Magnet Axial Field Flux-Switching Memory Machine. *IEEE Trans. Energy Convers.* **2018**, *33*, 2210–2219. [CrossRef]
20. Wardach, M.; Palka, R.; Paplicki, P.; Prajzendanc, P.; Zarebski, T. Modern Hybrid Excited Electric Machines. *Energies* **2020**, *13*, 5910. [CrossRef]
21. Zhang, G.; Hua, W.; Cheng, M.; Liao, J.; Wang, K.; Zhang, J. Investigation of an improved hybrid-excitation flux-switching brushless machine for HEV/EV applications. *IEEE Trans. Ind. Appl.* **2015**, *51*, 3791–3799. [CrossRef]
22. Pollock, C.; Pollock, H.; Barron, R.; Coles, J.R.; Moule, D.; Court, A.; Sutton, R. Flux-switching motors for automotive applications. *IEEE Trans. Ind. Appl.* **2006**, *42*, 1177–1184. [CrossRef]
23. Zhao, G.; Hua, W.; Qi, J. Comparative Study of Wound-Field Flux-Switching Machines and Switched Reluctance Machines. *IEEE Trans. Ind. Appl.* **2019**, *55*, 2581–2591. [CrossRef]
24. Zhu, Z.Q.; Wu, Z.Z.; Evans, D.J.; Chu, W.Q. A Wound Field Switched Flux Machine with Field and Armature Windings Separately Wound in Double Stators. *IEEE Trans. Energy Convers.* **2015**, *30*, 772–783. [CrossRef]
25. Yang, S.; Zhang, J.; Jiang, J. Modeling Torque Characteristics and Maximum Torque Control of a Three-Phase, DC-Excited Flux-Switching Machine. *IEEE Trans. Magn.* **2016**, *52*, 1–4. [CrossRef]
26. Nguyen, H.Q.; Jiang, J.Y.; Yang, S.M. Design of a wound-field flux switching machine with dual-stator to reduce unbalanced shaft magnetic force. *J. Chin. Inst. Eng.* **2017**, *40*, 441–448. [CrossRef]
27. Parappagoudar, M.B.; Pratihar, D.K.; Datta, G.L. Forward and reverse mappings in green sand mould system using neural networks. *Appl. Soft Comput.* **2008**, *8*, 239–260. [CrossRef]
28. Zhao, J.; Guan, X.; Li, C.; Mou, Q.; Chen, Z. Comprehensive evaluation of inter-turn short circuit faults in pmsm used for electric vehicles. *IEEE Trans. Intell. Transp. Syst.* **2020**, *22*, 611–621. [CrossRef]

energies

MDPI

Article

Hybrid Switched Reluctance Motors for Electric Vehicle Applications with High Torque Capability without Permanent Magnet

Vijina Abhijith [1,*], M. J. Hossain [1,*], Gang Lei [1], Premlal Ajikumar Sreelekha [2,*], Tibinmon Pulimoottil Monichan [2] and Sree Venkateswara Rao [3]

[1] School of Electrical and Data Engineering, University of Technology Sydney, Ultimo, NSW 2007, Australia
[2] Entuple E-Mobility, Diamond District, HAL Old Airport Road, Kodihalli, Bangaluru 560008, India
[3] Gyrate the Motor Inside, Shimla Block, Hill country, Bachupallay, Telangana 500090, India
* Correspondence: vijina.abhijith@student.uts.edu.au (V.A.); jahangir.hossain@uts.edu.au (M.J.H.); premlal95111@gmail.com (P.A.S.)

Abstract: In electric vehicle (EV) applications, hybrid excitation of switched reluctance motors (HESRMs) are gaining popularity due to their advantages over other EV motors. The benefits include control flexibility, simple construction, high torque/power density, and the ability to operate over a broad speed range. However, modern HESRMs are constructed by increasing the air gap flux density with permanent magnets (PMs) in the excitation system in order to generate more electromagnetic torque. This study aims to investigate a new topology for increasing the torque capabilities of HESRM without the use of permanent magnets (PMs) or other rare-earth components. This paper provides a comprehensive evaluation of the static and dynamic characteristics, software analysis using the Ansys 2D finite element method (FEM), and an experimental demonstration of the real-time motor with an advanced control strategy in MATLAB/Simulink. Our simulation and experimental results for a machine with 12/8 poles and a machine rating of 1.2 kW indicate that the HESRM designed without PMs has greater torque capability and efficiency than the conventional SRM. The proposed HESRM without PMs has a high torque/power density and a higher torque per ampere across the entire speed range, making it suitable for EV applications.

Keywords: switched reluctance motor (SRMs); hybrid excitation of SRM (HESRM); hybrid excitation of SRM without a permanent magnet; permanent magnet (PM); electric vehicle motor

Citation: Abhijith, V.; Hossain, M.J.; Lei, G.; Sreelekha, P.A.; Monichan, T.P.; Rao, S.V. Hybrid Switched Reluctance Motors for Electric Vehicle Applications with High Torque Capability without Permanent Magnet. *Energies* **2022**, *15*, 7931. https://doi.org/10.3390/en15217931

Academic Editors: Vladimir Prakht, Mohamed N. Ibrahim and Vadim Kazakbaev

Received: 24 September 2022
Accepted: 21 October 2022
Published: 26 October 2022

1. Introduction

The widespread use of ICE (internal combustion engine) automobiles has led to an increase in pollution and energy constraints in recent years. The use of electric vehicles (EVs) has the potential to reduce environmental stress, reduce pollution, and limit the number of vehicles requiring charging [1]. Electric vehicles (EVs) rely heavily on electric motors [2]; thus, numerous electric motors were evaluated and analyzed to determine which would be the most effective for various applications. These days, brushless DC motors (BLDCMs), induction motors, and different permanent magnet (PM) machines are some of the motor types other than SRM motors that can be used effectively in variable-speed applications [3,4]. In comparison with conventional motors, SRMs are appealing to researchers due to their low complexity, high durability, cost-effectiveness, and ease of production [5,6]. SRM appears to have the same power density as induction motors, and its reliability appears to be far superior to that of DC series and permanent magnet motors [7]. As the SRM lacks the mechanical components of other conventional motors, such as permanent magnets and rotor windings, it can be utilized in high-temperature environments [8]. Low torque and power density, compared with permanent magnet motors, as well as significant torque ripple, are two drawbacks that limit its utility [9].

Numerous researchers are devoted to reducing the machine architecture to the level required by applications in order to address these problems and boost the performance of SRMs. Prior research has demonstrated that numerous approaches to switched reluctance motors focus on innovative controller circuits [10–12], core segmentation [13–16], and hybrid excitations (HESRMs).

Using sophisticated controller circuits, the switching circuit of the stator of an SRM can be optimized to provide maximum torque. Precision phase-winding switching makes it possible to tailor the performance of modern electronic converters to specific applications [17]. This accurate switching method permits substantial control over the torque ripple of the motor. In order to increase the available flux in the air gap and the resulting flux, the segmentation method [18] is applied to either the stator or rotor cores. By limiting the flux to the separated regions, the average torque of the active region is increased. This segmented system diminishes mechanical strength and contributes to higher costs [19] due to the increased production complexity. Due to the internal flux's reciprocal coupling, the overall efficiency of the segmented core is diminished. Drive circuits with multiple stator segments may be used for low- and high-speed applications and provide more power and torque per ampere [20]. This technique increases torque output at the expense of machine durability [21]. The third option (HESRM) is permanent magnet hybrid-excited switched reluctance motors.

SRMs underwent a paradigm shift with the introduction of the hybrid excitation technique, which substantially increased their ability to generate electromagnetic torque. However, this approach uses PMs into the structure of the machine to increase the motor's average torque. A 12/8 HESRM with a modular stator was presented in a stator segmented with PMs [22]. There is evidence that it can transport a maximum flux density at a higher torque [23]. The flux capacity is increased by 34.7% compared with the segmented SRM without PM [24]. As a result, the average torque is 71.4% greater than the conventional SRM. A comparison study of a segmented stator permanent magnet hybrid excitation SRM drive [25] reported that the static average torque of the segmented HESRM is 1.16 to 1.39 times that of the standard one. There were six permanent magnets (PMs) in the stator of the machine; due to the behavior of the PMs, this could cause cogging torque. It has been demonstrated that SRM with auxiliary winding and permanent magnets is 3% more efficient and reduces size and weight by the same amount [26]. The HESRM, meanwhile, combines the benefits of the permanent magnet machine and the electromagnet machine to create a hybrid excitation of the double salient machine [27]. The rare-earth permanent magnets in the rotor core of the 6-slot 8-pole hybrid excitation flux switching machine (HEFSM) are an innovative design element that contributes to the machine's high power density by providing adequate mechanical strength with enhanced electromagnetic performances [28]. The suggested machine, the HEFSM, has an efficiency ranging from 92 to 97 percent, and it is also used to dampen vibration and acoustic noise. Similar to low-power, low-cost applications, single-phase hybrid SRMs are utilized for low-power, low-cost applications. The new low-cost hybrid SRM for variable-speed pumps contains permanent magnets in the shape of rectangular shards. As a result, both the average torque and the torque ripple are increased; see [29].

However, the hybrid excitation method improves the efficiency of SRM due to the strategic placement of PMs. The PM's actions have the potential to deceive the SRM because they alter the SRM's typical behavior. Additionally, rare-earth metals and PMs are becoming increasingly scarce. In the era of electric automobiles, both cost-effective and efficient motors will be crucial. Nonetheless, the high demand and limited supply of PMs will increase the cost of manufacturing PM-equipped machines. To maintain a high torque/power density under these conditions, the PM's electric motor integration strategy is not a viable option. Therefore, this work primarily contributes to eliminating the need for permanent magnets in SRMs, while simultaneously enhancing torque performance via hybrid excitation techniques.

Due to the combined action of the two controllers, the proposed HESRM motor has better control over the motor. Because separate controllers are used for the main and auxiliary windings, it has more flexibility. The torque per ampere is improved by the proposed method because it produces more torque at lower current levels than the traditional SRM. The auxiliary windings can work at current control, while the main windings can work in speed control. To increase torque, auxiliary windings can be used to inject DC, and thus maximize the machine's potential. There is less redundancy because the motor is switched by two separate controllers, one of which can continue to operate the machine even if the other fails. This machine exhibits a significantly higher flux in its stator yoke than conventional machines, which is primarily caused by flux saturation in the core.

The remainder of this paper is organized as follows: Section 2 provides a detailed description of the proposed HESRM's machine topology and fundamental operations. In Section 3, the simulation results of the proposed machine's static and dynamic behavior, including the results of the machine's flux distribution, phase flux linkage, electromagnetic torque, etc., are described in detail. Furthermore, Section 4 discusses the construction of machines with identical dimensions and experimental validation. In Section 5, the average torques of HESRM are compared with those of a conventional SRM. Finally, Sections 6 and 7 cover the discussion and conclusion in detail.

2. Machine Topology for the Proposed Machine

The machine topologies of conventional SRM and the proposed HESRM without PMs are shown in Figure 1. Both motors have twelve stator and eight rotor poles without any PMs incorporated within the cores. Still, the windings of poles per phase are assigned for different excitation—main pole windings and auxiliary winding excitations, respectively, for the assigned poles. The new hybrid excitation technique's main goal is to boost the conventional SRM's torque performance without sacrificing any of its distinctive characteristics. Permanent magnets and structural changes are entirely avoided in HESRM. The excitation of the conventional SRM differs in that the windings are excited and separated separately to inject more external current into the HESRM. These auxiliary windings are selected so that two windings from each phase are separated, maintaining the motor's mechanical and electrical symmetry. Figure 1 depicts these alternately positioned auxiliary windings, which inject external DC current in accordance with the torque required by the controller. The motors' measured speed is sensed and fed back to create PWMs, which are then delivered to the throttle.

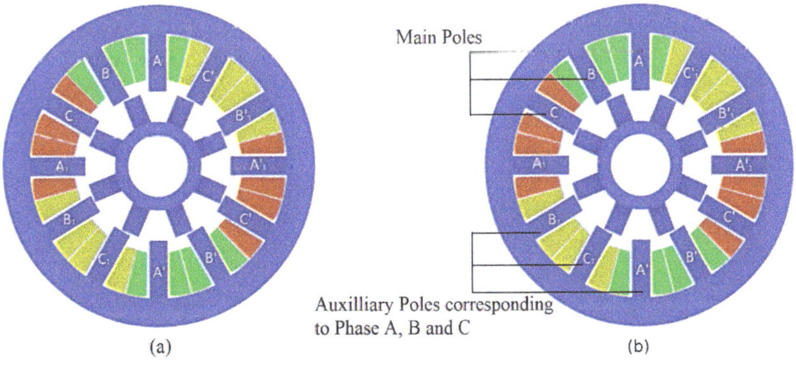

Figure 1. Machine topologies of two SRMs with the same rating as shown as (**a**) 12/8-pole conventional SRM and (**b**) 12/8-pole HESRM without PMs.

First, the poles are assigned for auxiliary excitation. The stator of the HESRM differs from that of the conventional SRM; in the HESRM, the stator poles are assigned to the main and auxiliary poles based on the excitation. Since there are twelve stator poles in the motor, six poles can be used for the primary excitation, while the remaining poles can be used for the auxiliary excitation to provide the same three phases as conventional SRMs. Second, the number of poles for the primary and auxiliary windings is maintained constant in order to inject external flux into the poles simultaneously with the excitation of the primary windings. Separate DC excitation can increase the resultant flux in the air gap due to the symmetrical nature of the poles. Using electrical and mechanical symmetry, the primary and secondary poles are differentiated. This geometry of twelve stators can be utilized effectively to provide two distinct windings for each phase [30]. There are two distinct windings in HESRM: the main windings and the auxiliary windings. With the aid of a separately excited driver circuit, the auxiliary windings' poles inject DC current. The main windings provide the current for excitation of the motor. Because the variations in both windings' currents affect the air-gap flux density, the flux in HESRM is more than the traditional SRM. The new topology has a greater flux density, power, and torque without using the PM behavior.

Figure 2 is a structural diagram of magnetic field distribution that illustrates the fundamental operating principle of the proposed HESRM without PMs. It shows the magnetic field distribution of two SRMs with aligned and misaligned poles. Regardless of whether the SRMs are operating in an aligned or unaligned position, identical field lines can be observed. The HESRM flux is injected along the same path as the carryover path for the primary winding flux. Therefore, the flux due to auxiliary excitation can carry a greater flux to generate a greater electromagnetic torque. A detailed block diagram of the HESRM is shown in Figure 3. The new hybrid excitation technique's main goal is to boost the conventional SRM's torque performance without sacrificing any of its distinctive characteristics. Permanent magnets and structural changes are entirely avoided in HESRM. The excitation of the conventional SRM differs in that the windings are excited and separated separately to inject more external current into the HESRM. These auxiliary windings are selected so that two windings from each phase are separated, maintaining the motor's mechanical and electrical symmetry.

The net flux generated by HESRM in the absence of PMs travels through the core and then the air gap, creating a closed path similar to the conventional flux path. When both coils are excited, the flux intensity generated by the primary and secondary differs because the secondary pole can carry a greater current than the primary. The total flux produced by the primary coil and the auxiliary coils can be observed to be added. Consequently, the total magnetic flux produced by the HESRM is greater than the conventional flux without PMs. Consequently, the electromagnetic torque results from the algebraic sum of the total effective flux of both windings.

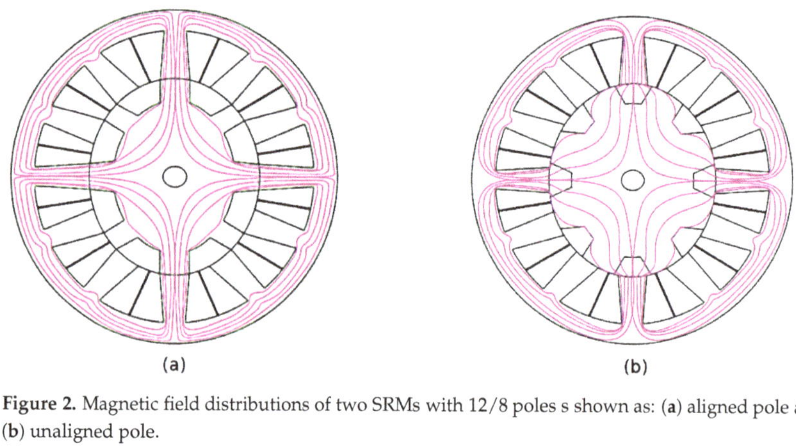

Figure 2. Magnetic field distributions of two SRMs with 12/8 poles s shown as: (**a**) aligned pole and (**b**) unaligned pole.

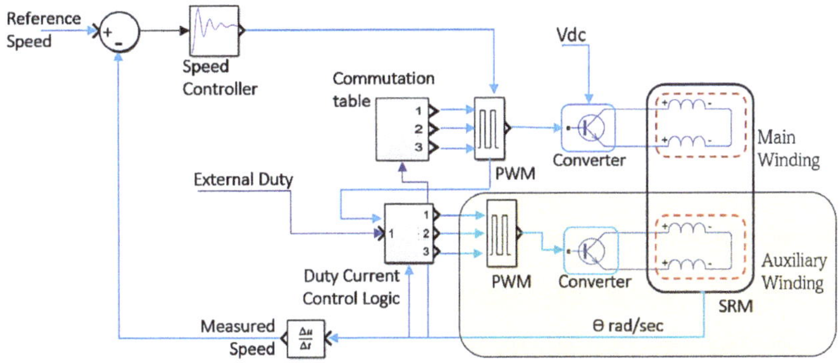

Figure 3. Block diagram of the proposed HESRM without permanent magnet excitation.

The dimensions and specifications of the conventional SRM and the proposed HESRM are listed in Table 1. Similarly, the number of phases, number of poles in the stator and rotor, rated power, speed, length of the core, and type of material are all identical. In addition, the assignment of the HESRM's auxiliary and primary poles differs based on the flux path.

Table 1. Main dimensions and parameters of two SRMs.

Dimensions	Conventional SRM	HESRM
Phase number	3	3
Stator and rotor poles	12/8	12/8
Hybrid excitation poles	12	6-6
Rated Power(kW)	1.2	1.2
Rated speed (rpm)	2500	2500
Outer dia of stator (mm)	138	138
Inner dia of rotor (mm)	81.50	81.50
Length of air gap (mm)	0.5	0.5
Type of material	M15_26G	M15_26G

3. Characteristics of the Proposed HESRM Using 2D FEA Analysis

This section presents the magnetic characteristics of the conventional SRM at aligned and unaligned positions. To obtain better clarity and easiness of comparison, both the SRMs are given the same size, number of coils per pole, and similar types of windings, as shown in Table 1. The field lines where the stator and rotor iron have the shortest flux path are concentrated in the aligned position rather than the unaligned one.

3.1. Magnetic Characteristics of Two SRM's

The cross-sectional areas of the two SRMs are the same due to the same rotor construction and geometric dimensions. Thus, both SRMs, the stators, and rotor iron weights identical to the constructions have a 7.933 (kg) total net weight, and the stator and rotor core steel consumptions (kg) are 7.05683 and 2.6735, respectively. Owing to the difference, only the excitation scheme provided the winding with different currents.

Field Line and Flux Distributions

The magnetic characteristics of the proposed SRM are similar to that of the conventional SRM. Here, the quantity of the flux lines will increase as more current is injected into the auxiliary poles. As a result, the effective flux lines in the air gap increase, and the resultant electromagnetic torque increases. Figures 4 and 5 demonstrate the magnetic flux distribution and the flux density of the conventional SRM and the proposed HESRM, respectively. The excitation for the three-phase scheme for the conventional SRM with aligned and unaligned poles shows that the minimum flux density is 1.2–1.3 Tesla (T) for the unaligned pole and 1.7–1.8 T for aligned poles. The conventional SRM's range for maximum flux density is 0–1.9 T, and steel is the material used for the simulation. Figure 6 shows the conventional SRM for the given excitation (5A) at aligned and unaligned positions. At aligned position, it is 1.2 T, and at unaligned position, it is 1.6 T. Because the auxiliary windings' unexcited pole is aligned with the excitation pole in the HESRM, it can reach 1.06 T in alignment with the same excitation current (5A). Auxiliary injection current helps to increase the magnetic flux density in an unaligned state in 1.3 T, comparing the magnetic flux distribution and flux intensity of the proposed HESRM, where both characteristics are identical. Accordingly, there is only one difference in the intensity of flux carrying in the air gap due to different DC injections in the auxiliary poles. First, the excited state produces the main flux with the effect of the main winding, and the extra DC provided by the auxiliary excitation creates more electromagnetic torque.

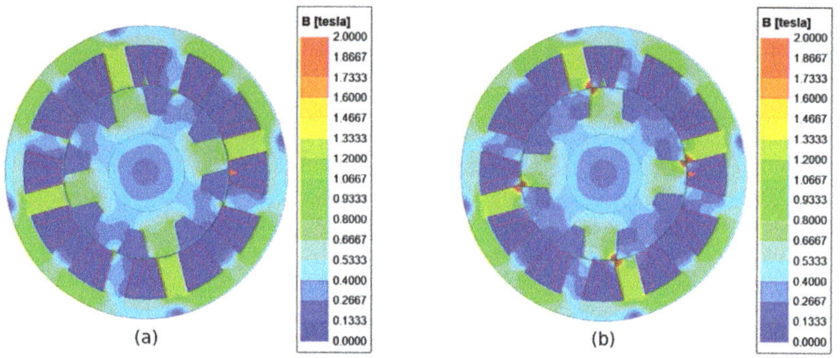

Figure 4. Magnetic flux distribution of conventional SRM: (**a**) aligned poles and (**b**) unaligned poles.

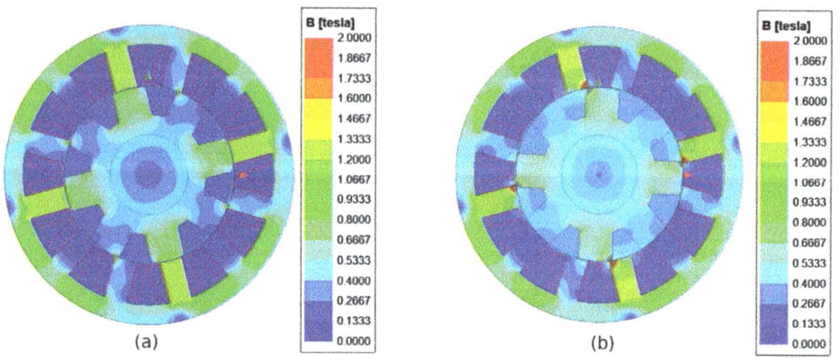

Figure 5. Magnetic flux distribution of HESRM: (**a**) aligned poles and (**b**) unaligned poles.

The reluctance principle governs SRM. Therefore, the SRM generates the torque based on the magnetic reluctance of the motor. The magnetic circuit between the rotor and stator is very reluctant when they are not in alignment. As the rotor tries to align with the powered stator poles at this point, the stator pole pairs are activated, reducing the magnetic reluctance. Reluctance torque is formed when the rotor is able to reach the minimal point of reluctance. It is necessary to precisely time the stator poles' excitation so that it only happens when the rotor is attempting to align with the exciting pole. SRM may require positive feedback from encoders or Hall effect sensors in order to control the excitation of the stator based on a precise rotor position.

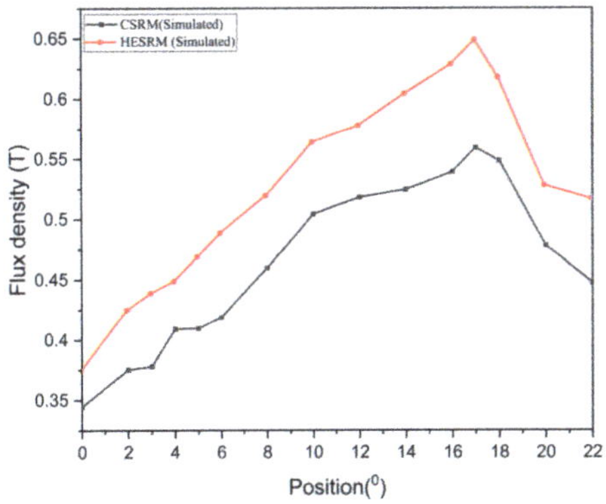

Figure 6. Simulated flux density of conventional SRM and the hybrid DC-excited SRM.

3.2. Steady-State Characteristics

The electromagnetic behaviour of the proposed HESRM was analyzed through the FEM (finite element method) using commercial software ANSYS/Maxwell. The motor drive simulation was carried out by MATLAB/Simulink, and the interpretation was held by the co-simulation method. For this purpose, the feasibility of the proof of concept was verified with the help of software simulations by analyzing steady-state and dynamic behaviors.

Static Electromagnetic Torque

The static magnetic torque characteristics of two SRMs with phase excitations are shown in Figure 7. The conventional SRM static torque curves with the current of 15 A vary from 0 to 22.5 degrees, as shown in Figure 7a. Accordingly, the static torque for one-half of the excitation cycle of conventional SRM attains unalignment to the aligned pole. It can be seen that the torque varies from minimum to maximum and finally reaches a minimum with the unaligned position. The maximum static torque that produces the phase excitation current of 15 A is 2 N m. From the comparison, it is clear that the auxiliary excitation in which a different current is provided with the help of DC injection may attain different patterns of static currents. The auxiliary winding has a provision for carrying the maximum current to the rated value. Figure 7b shows that the static flux curve pattern increases the effect of varying according to the rises of DC in the auxiliary coils. Since additional current is passed through DC injection in auxiliary windings using special circuitry, the flux inside the core increases, which results in higher torque and power density.

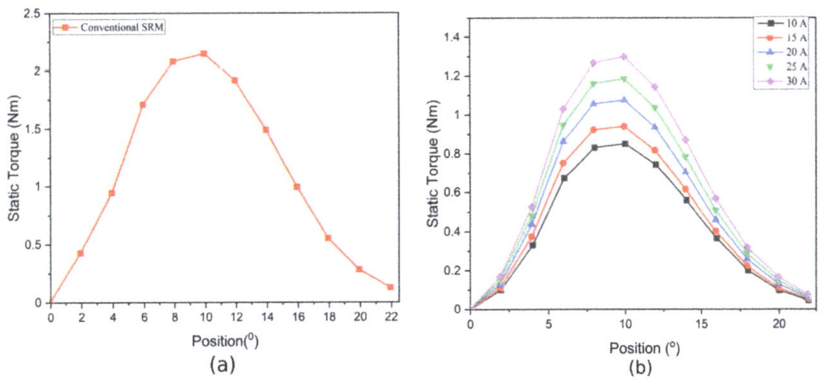

Figure 7. Static torque characteristics: (**a**) conventional SRM at rated current 15 A. (**b**) HESRM under various currents with respect to the DC field injection.

3.3. Dynamic Performance of Two SRM Drives

MATLAB/Simulink was used to study the performance of the motor drive system for two SRMs. The proposed HESRM drive system's Simulink model has an asymmetric half-bridge converter, logic PWM, angle controller mechanical systems, DC sources, speed and current controllers, a separate auxiliary excitation with an extra switch, etc. The machine model is developed on the Ansys/Maxwell workbench. The dynamic performance of both machines was performed and compared under the same conditions.

3.3.1. Gate Pulse Characteristics of Two SRMs

Figure 8 shows the moving torque and current characteristics of two SRMs in which the gate pulses are given to the phases at the same turn ON and turn OFF times. In conventional SRM, the single PWM pulses sequentially provide the switching scheme for all phases. The hybrid excitation method provides identical gate pulses to the main and auxiliary windings separately. The torque characteristics of conventional SRM can be achieved with similar excitations in HESRM. Since the moving torque remains similar, the auxiliary current can be further increased to the rated value to achieve maximum performance.

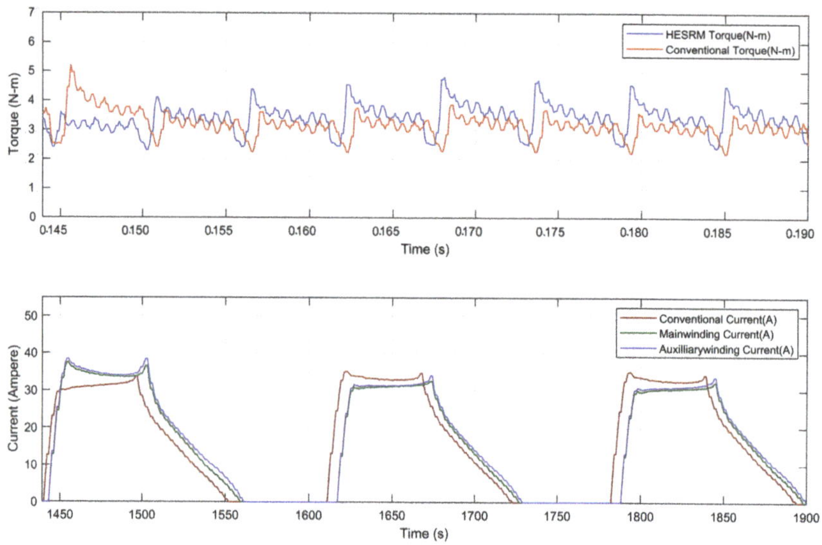

Figure 8. Comparison of characteristics of moving torque and current of two SRMs.

3.3.2. Closed-Loop Current Control of Two SRMs

The closed-loop current control of the conventional SRM is established by obtaining current feedback from the windings and providing input to the current controller. The conventional and proposed HESRM showed similar results at various input currents with matching moving torque and slightly less ripple, as shown in Figure 9. The current from the windings can be sensed and used to provide additional flux through the auxiliary in this hybrid excitation method.

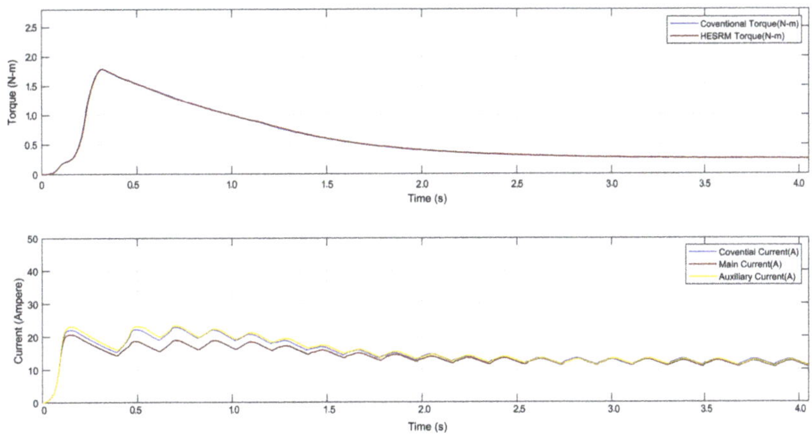

Figure 9. Comparison of closed-loop current control of conventional SRMs and HESRM.

3.3.3. Variable-Speed Control Strategy for Two SRMs

In Figure 10, the speed control is established using a proportional–integral (PI) controller by measuring the actual speed and comparing it with the reference speed, which varies continuously from minimum to rated value. The proportional (k_p) and integral (k_i) values of the controller are tuned in such a way as to reach the set points as early as possible. The analysis of the two results shows similar characteristics at the same current rating and load. The additional torque needed for the motor can be identified and provided through these auxiliary windings to obtain better results. However, the HESRM method is derived to enhance the performance by separately injecting the DC with the help of auxiliary winding. Hence, these results elaborate on how to achieve the same performance through the different winding excitations separately. However, the HESRM makes it possible only through different currents passing through the assigned winding. This method benefits by changing different currents through the controller to improve the average torque characteristics and reduce the torque ripple to the minimum level.

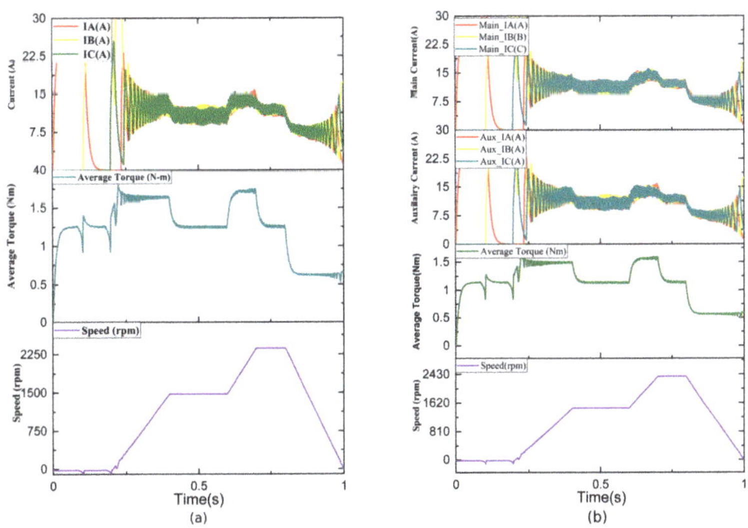

Figure 10. Closed-loop variable-speed current control strategy: (**a**) conventional SRM and (**b**) HESRM; the curves are current (A), torque (N-m), and speed (rpm).

In a conventional SRM, the variable speed is related to the main current excitation and torque, whereas in a HESRM, where there are two currents—the main current and an auxiliary current—the main current is responsible for both the excitations, and the auxiliary speed current is responsible for the variable speed. Figure 10 displays the torque and current characteristics of the traditional and hybrid SRM at variable speeds. To simulate how EVs would operate in real time, the speed input is changed using ramps and steps. Steps of 1000 rpm are used to ramp up the speed from zero to 2500 rpm. The graphs show all three phases of the currents at various speeds. Since this motor is a three-phase switching machine, the irregularities in the phase currents are caused by the speed changes. The motor's transient behavior is indicated by the longer switching patterns at the start of the operation.

4. Experimental Results

4.1. Motor Prototype

The torque characteristic of the SRM is the most important performance parameter that must be measured. As shown in Figure 11, in order to analyze the characteristics, a back-to-back motor setup with a torque sensor coupled in between was constructed. The motor used on the load side is a Permanent Magnet Synchronous Machine (PMSM) rated at 5000 revolutions per minute (rpm). This motor is tightly coupled with the motor of the plant. The non-linearity of the Futek torque sensor (model: TRD605) is +/−0.2%. This high-precision strain gauge sensor measures up to 250 N-m. The load-side PMSM contains an incremental speed encoder with a pulse per revolution (PPR) of 1024, which can be increased to 5000 by counting the rising and falling edges of the encoder's pulses. The PMSM's back EMF is used to drive a three-phase rectifier to a resistive load. Changes to the resistance can be made to load the motor.

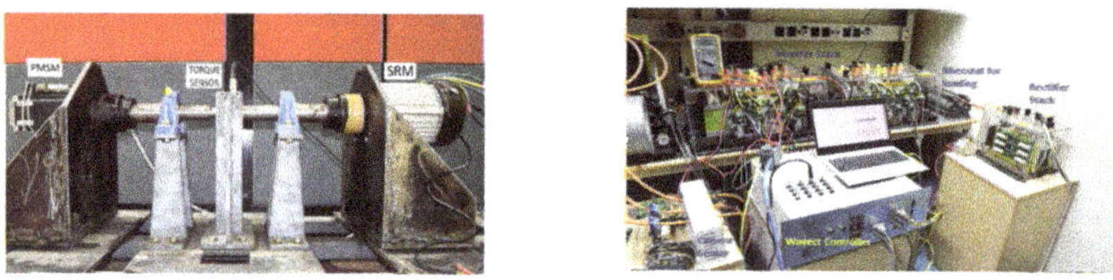

(a) (b)

Figure 11. (**a**) Experimental setup of static measurements and (**b**) controller test setup of the proposed HESRM.

4.2. Dynamic Performance of HESRM without PMs

The dynamic performance of the conventional SRM and the proposed HESRM can be analyzed by enabling a closed-loop model. The experimental setup for both SRM drives is identical in all aspects and conditions. For the analysis, conventional drive systems are used, while in the auxiliary, additional circuits provide excitation. The motor has three hall sensors that are electrically separated by one degree. Using an asymmetric inverter stack, the three phases can be changed to run the motor. The control mechanism was enabled using Wavect, a universal rapid control prototyping platform for Motor Control Drives developed by Entuple Technologies, India. This FPGA-based controller implementation was developed to analyze real-time strategies that can be modeled, tested, and managed. The current sensors in the "wavect controller" can obtain the instantaneous phase currents, and torque can be compared via an ADC channel in the controller. The primary advantage of this mechanism is the ability to compare current and torque simultaneously. Figure 12 depicts the rewinding structure and the waveform corresponding to the obtained moving torque and peak current when the motor is tested at various speeds and loaded to its peak current.

Figure 12. Photograph of rewinding the conventional SRM to HESRM (a) Main winding (b) Auxilary winding.

Figure 13 depicts the conventional SRM phase current and torque performance at 1000 and 2000 rpm. Figure 14 depicts the analysis of the auxiliary current and torque for the hybrid excitation method. The three-phase current is denoted by different colours, and green indicates the auxiliary exciting current. In this method, both the primary and secondary poles are excited simultaneously without injecting any currents; the performance matches the Ansys/Matlab software simulation results.

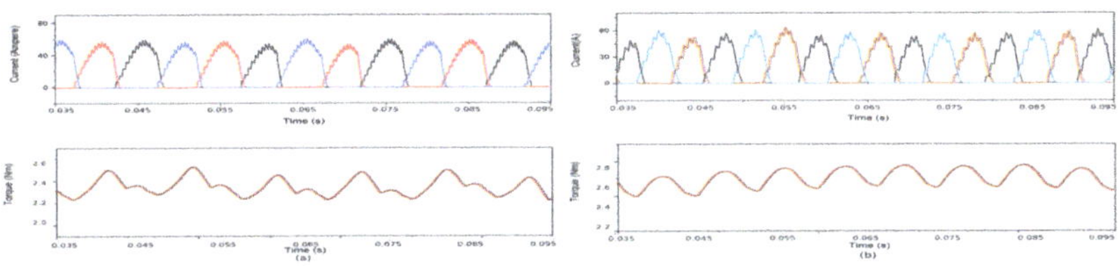

Figure 13. Conventional SRM dynamic performance of variable speed: (**a**) 1000 rpm and (**b**) 2500 rpm.

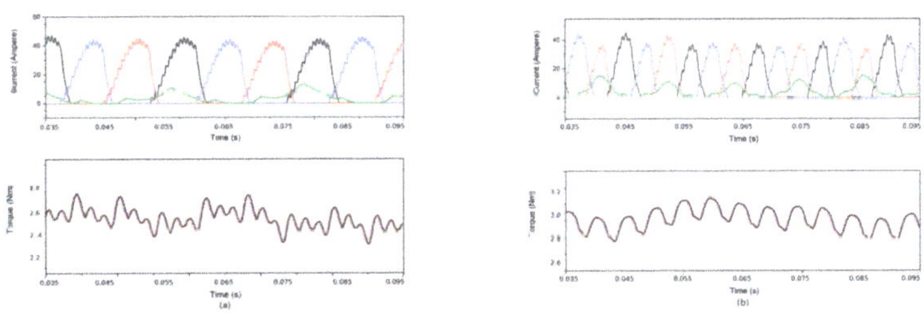

Figure 14. Proposed HESRM dynamic performance of variable speed: (**a**) 1000 rpm and (**b**) 2500 rpm.

The transient start-up response of the torque, phase current, and speed for two SRM drives with closed-loop control are tested and shown in Figure 15. The results under the same condition were evaluated with the DC-link voltage at 60 V and the load torque at 2 N m, respectively. It was found that both SRMs perform start-up in a similar way, according to the switching scheme provided, and accelerate rapidly as per the commanded speed of 1000 rpm to rated 2500 rpm. They reach the time of 50 ms and 500 ms, respectively. In other words, the speed changes from 500 rpm to 2500 rpm under closed-loop control, and the gain parameter changes in a real-time manner to reach the commanded speed. The result shows the rotor speed tracks the command signal in all manner and achieves the substantial results obtained from the simulated one.

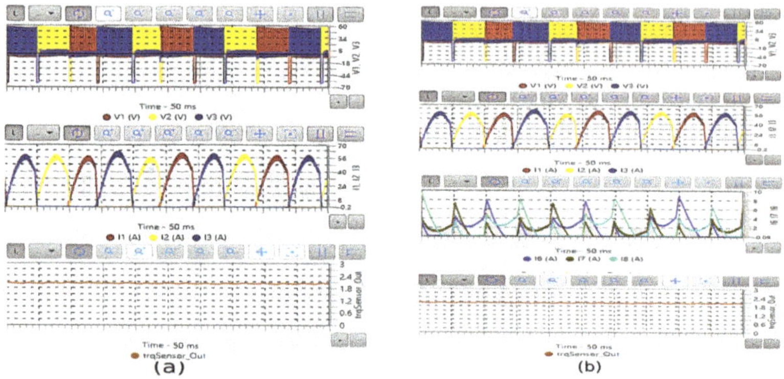

Figure 15. Closed-loop transient response at reference speed: (**a**) conventional SRM and (**b**) HESRM.

Figure 16 depicts the comparison between the measured and simulated average torque characteristics. The graphical analysis shows that the measured and simulated torque values are nearly identical. The average torque increased between 1000 and 2500 revolutions per minute. In Figure 16a, at an initial speed of 1000 rpm, the average torque begins at 1.8 N m. It increases exponentially as the effect of the auxiliary current varies. Variable-speed motors are permitted to adjust the auxiliary current from minimum to maximum, up to the rated value. Therefore, there is a tendency to provide more flux, thereby increasing the probability that the electromagnetic torque in the active region will increase.

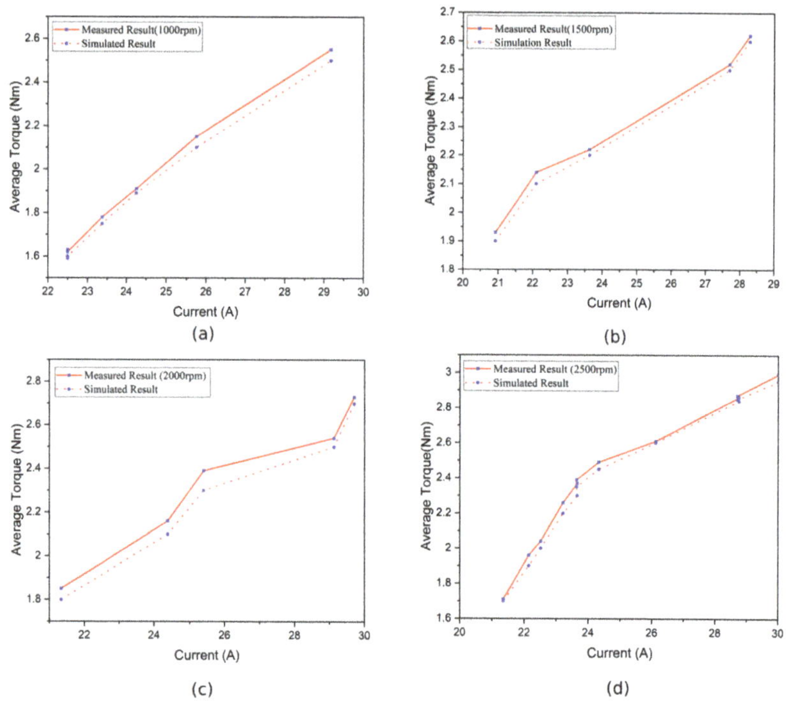

Figure 16. Average torque performance for variable speed of proposed HESRM: (**a**) 1000, (**b**) 1500, (**c**) 2000, and (**d**) 2500.

5. Comparison of Hardware Solution with FEM Simulation Results

In order to validate the proof of concept with results from the software analysis and theoretical perdition, a 12/8 conventional SRM and a rewound SRM of the same size for the proposed method were prototyped and experimentally validated. The main dimensions and parameter specifications of the two SRMs are shown in Table 1.

Figure 17 depicts the comparison of the average torque between the conventional SRM and the proposed HESRM at various speeds. To demonstrate the benefit of this novel machine structure, the hybrid excitation without a permanent magnet is compared with two SRMs from the same platform. The average performances for the speed range vary from 1000 rpm to 2500 rpm, shown in different figures. At 1000 rpm, the torque is 2 N m with the conventional current and excitation, while it reaches 2.5 N m with the auxiliary. For this reason, there is a significant increase in torque performance, and the percentage of torque increase at all speeds is greater than 20%. The pattern of torque response increase varies depending on speed and load.

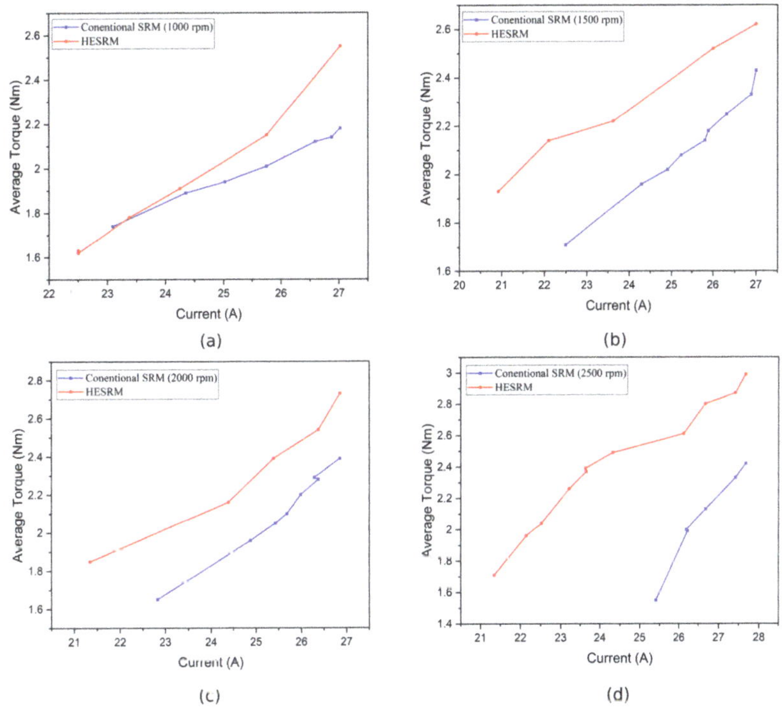

Figure 17. Comparison of average torque with variable speed for conventional SRM and proposed HESRM: (**a**) 1000, (**b**) 1500, (**c**) 2000, and (**d**) 2500 rpm.

6. Discussion

Figure 18a shows the plots of conventional and HESRM's average torque characteristics for various speeds. Both have a difference of 1 to 2 N m at 500 rpm, and it gradually increases. The HESRM reaches almost 2.4 N m at 1500 rpm, whereas the conventional one is only 2.1 N m. The HESRM reaches a maximum torque of 2.6 N m at 2500 rpm with a lead of almost 0.3 N m.

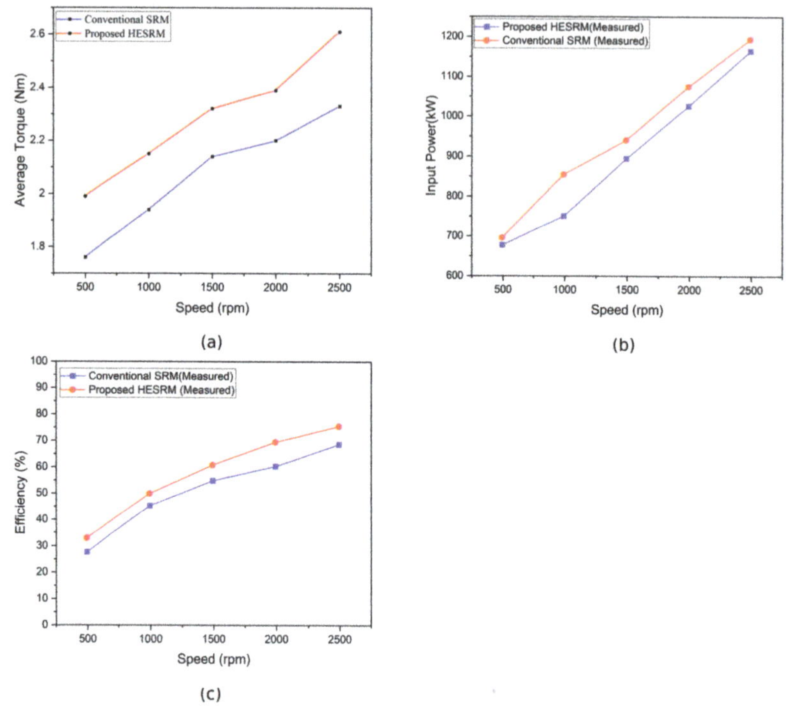

Figure 18. Comparison of conventional SRM with proposed HESRM: (**a**) average torque, (**b**) efficiency, and (**c**) input power.

The efficiencies for both SRMs were almost identical at the beginning of the test at 500 rpm. As speed increases, the efficiency of the HESRM makes clear leads compared with the conventional one. Up to 1500 rpm, the efficiencies of both motors follow a linear path with a difference of approximately 10%. At the rated speed of 2500 rpm, the efficiency of the motor differs to a maximum value of 80% and a minimum of 65%. The efficiency of the HESRM increased by 20%, as shown in Figure 18b.

The input power consumed by both motors is almost constant at 500 rpm, as clearly seen in Figure 18c. As the speed increases, the input power consumed by the motors starts to increase. The conventional SRM intakes more power at all the respective rpm compared with the proposed one. At 1000 rpm, the conventional SRM draws almost 850 W, whereas the HESRM draws a minimum of 700 W. At maximum speed, the input power consumed has a difference of almost 50 W.

Torque ripple is one of the major drawbacks of SRM, and thereby the hybrid excitation methods tend to provide higher possibilities. The proposed new HESRM without PM tends to exhibit less torque ripple than the conventional HESRM. In contrast, the conventional SRM exhibits a ripple of 3.72% at 2500 rpm, which decreases to 3.69% at 2000 rpm. Simultaneously, the proposed HESRM reduces torque ripple as a result of reduced pole switching, achieving 1.44% and 1.69% at 2500 and 2000 rpm, respectively.

7. Conclusions

When the average, static, and speed–torque characteristics of conventional and DC-injected hybrid SRMs are compared, it was found that significant improvements were made by HESRM over conventional SRM. In addition, the properties and performance of the machines are improved without sacrificing their fundamental qualities. In addition to a high torque-per-ampere rating, it offers high efficiency, high dependability, and redundancy.

The standard SRM is outperformed by the proposed method, and it generates greater torque at significantly lower current levels. Furthermore, the proposed HESRM employs separate excitations for the windings, thereby enhancing the versatility of the controller.

The proposed topology may be a replacement for HESRM machines with PM insertion. Since this study clearly demonstrates enhanced performance, the proposed system can be implemented in a variety of applications, such as electric vehicles. Moreover, improved average and smooth speed–torque characteristics have been achieved, which can be an excellent solution for EVs. Finally, this new topology can achieve greater efficiency with reduced torque ripple. Future objectives of this research include the design and implementation of advanced, optimized controllers to further reduce torque ripples. We are planning to study variable-speed control strategies for SRMs along with new designs.

Author Contributions: Conceptualization, V.A.; formal analysis, V.A. and M.J.H.; investigation, V.A. and P.A.S.; writing—original draft preparation, V.A.; writing—review and editing, M.J.H. and G.L.; supervision, M.J.H. and G.L.; industrial support, S.V.R. and T.P.M., concept analysis M.J.H. and S.V.R.; experimental analysis, V.A. and P.A.S. All authors have read and agreed to the published version of the manuscript.

Funding: The research is funded by the "University of Technology, Sydney".

Data Availability Statement: Not applicable.

Acknowledgments: I would like to acknowledge "Entuple Technologies" for providing the universal controller 'Wavect 300' and "Entuple E Mobility", Bangalore, India, for supporting the entire hardware implementation and testing setup during this work.

Conflicts of Interest: The authors declare no conflict of interest.

Abbreviations

The following abbreviations are used in this manuscript:

SRM	Switched Reluctance Motor
HESRM	Hybrid Excitation of Switched Reluctance Motors
PM	Permanent Magnet
EVs	Electric Vehicles
BLDCMs	Brushless DC Machines
ICE	Internal Combustion Engine
FEM	Finite Element Analysis

References

1. Lan, Y.; Benomar, Y.; Deepak, K.; Aksoz, A.; Baghdadi, M.E. Switched Reluctance Motors and Drive Systems for Electric Vehicle Powertrains: State of the Art Analysis andFuture Trends. *Energies* **2021**, *14*, 2079. [CrossRef]
2. Xu, J.; Zhang, L.; Meng, D.; Su, H. Simulation, Verification and Optimization Design of Electromagnetic Vibration and Noise of Permanent Magnet Synchronous Motor for Vehicle. *Energies* **2022**, *15*, 5808. [CrossRef]
3. Men, X.; Guo, Y.; Wu, G.; Chen, S.; Shi, C. Implementation of an Improved Motor Control for Electric Vehicles. *Energies* **2022**, *15*, 4833. [CrossRef]
4. Martellucci, L.; Capata, R. High Performance Hybrid Vehicle Concept—Preliminary Study and Vehicle Packaging. *Energies* **2022**, *15*, 4025. [CrossRef]
5. Tomczewski, K.; Wrobel, K.; Rataj, D.; Trzmiel, G. A Switched Reluctance Motor Drive Controller Based on an FPGA Device with a Complex PID Regulator. *Energies* **2021**, *14*, 1423. [CrossRef]
6. Rahman, M.S.; Lukman, G.F.; Hieu, P.T.; Jeong, K.; Ahn, J. Optimization and Characteristics Analysis of High Torque Density 12/8 Switched Reluctance Motor Using Metaheuristic Gray Wolf Optimization Algorithm. *Energies* **2021**, *14*, 2013. [CrossRef]
7. Pupadubsin, R.; Kachapornkul, S.; Jitkreeyarn, P.; Somsiri, P.; Tungpimolrut, K. Investigation of Torque Performance and Flux Reversal Reduction of a Three-Phase 12/8 Switched Reluctance Motor Based on Winding Arrangement. *Energies* **2022**, *15*, 284. [CrossRef]
8. Wang, J.; Jiang, W.; Wang, S.; Lu, J.; Williams, B.W.; Wang, Q. A Novel Step Current Excitation Control Method to Reduce the Torque Ripple of Outer-Rotor Switched Reluctance Motors. *Energies* **2022**, *15*, 2852. [CrossRef]
9. Kocan, S.; Rafajdus, P.; Bastovansky, R.; Lenhard, R.; Stano, M. Design and Optimization of a High-Speed Switched Reluctance Motor. *Energies* **2021**, *14*, 6733. [CrossRef]

10. Ding, W.; Member; Liu, G.; Li, P. A hybrid control strategy of hybrid-excitation switched reluctance motor for torque ripple reduction and constant power extension. *IEEE Trans. Ind. Electron.* **2020**, *67*, 38–48. [CrossRef]
11. Kabir, M.A.; Husain, I. Design of Mutually Coupled Switched Reluctance Motors (MCSRMs) for Extended Speed Applications Using 3-Phase Standard Inverters. *IEEE Trans. Energy Convers.* **2016**, *31*, 436–445. [CrossRef]
12. Elmutalab, M.A.; Elrayyah, A.; Husain, T.; Sozer, Y. Extending the Speed Range of a Switched Reluctance Motor Using a Fast Demagnetizing Technique. *IEEE Trans. Ind. Appl.* **2018**, *54*, 3294–3304. [CrossRef]
13. Lee, C.; Krishnan, R.; Lobo, N.S. Novel Two-Phase Switched Reluctance Machine Using Common-Pole E-Core Structure: Concept, Analysis, and Experimental Verification. *IEEE Trans. Ind. Appl.* **2009**, *45*, 703–711. [CrossRef]
14. Mecrow, B.C.; El-Kharashi, E.A.; Finch, J.W.; Jack, A.G. Segmental rotor switched reluctance motors with single-tooth windings. *IEE-Electr. Power Appl.* **2003**, *150*, 591–599. [CrossRef]
15. Mecrow, B.C.; El-Kharashi, E.A.; Finch, J.W.; Jack, A.G. Preliminary Performance Evaluation of Switched Reluctance Motors With Segmental Rotors. *IEEE Trans. Energy Convers.* **2004**, *19*, 679–686. [CrossRef]
16. R, V.; Fernandes, B.G. Design Methodology for High-Performance Segmented Rotor Switched Reluctance Motors. *IEEE Trans. Energy Convers.* **2015**, *30*, 11–21. [CrossRef]
17. Ding, W.; Yang, S.; Hu, Y.; Li, S.; Wang, T.; Yin, Z. Design Consideration and Evaluation of a 12/8 High-Torque Modular-Stator Hybrid Excitation Switched Reluctance Machine for EV Applications. *IEEE Trans. Ind. Electron.* **2017**, *64*, 9221–9232. [CrossRef]
18. Ding, W.; Fu, H.; Hu, Y. Characteristics Assessment and Comparative Study of a Segmented-Stator Permanent-Magnet Hybrid-Excitation SRM Drive With High-Torque Capability. *IEEE Trans. Power Electron.* **2018**, *33*, 482–500. [CrossRef]
19. Hasegawa, Y.; Nakamura, K.; Ichinokura, O. A Novel Switched Reluctance Motor With the Auxiliary Windings and Permanent Magnets. *IEEE Trans. Magn.* **2012**, *48*, 3855–3858. [CrossRef]
20. Zhihui, C.; Yaping, S.; Yangguang, Y. Static Characteristics of a Novel Hybrid Excitation Doubly Salient Machine. In Proceedings of the International Conference on Electrical Machines and Systems (ICEMS), Nanjing, China, 27–29 September 2005. [CrossRef]
21. Sulaiman, E.; Kosaka, T.; Matsui, N. High Power Density Design of 6-Slot–8-Pole Hybrid Excitation Flux Switching Machine for Hybrid Electric Vehicles. *IEEE Trans. Magn.* **2011**, *47*, 4453–4456. [CrossRef]
22. Jia, S.; Qu, R.; Li, J.; Li, D.; Kong, W. A Stator-PM Consequent-Pole Vernier Machine With Hybrid Excitation and DC-Biased Sinusoidal Current. *IEEE Trans. Magn.* **2017**, *53*, 1–4. [CrossRef]
23. Ahn, J.-W.; Park, S.-J.; Lee, D.-H. Hybrid Excitation of SRM for Reduction of Vibration and Acoustic Noise. *IEEE Trans. Ind. Electron.* **2004**, *51*, 374–380. [CrossRef]
24. Lu, K.; Jakobsen, U.; Rasmussen, P.O. Single-Phase Hybrid Switched Reluctance Motor for Low-Power Low-Cost Applications. *IEEE Trans. Magn.* **2011**, *47*, 3288–3291. [CrossRef]
25. Lu, K.; Rasmussen, P.O.; Watkins, S.J.; Blaabjerg, F. A New Low-Cost Hybrid Switched Reluctance Motor for Adjustable-Speed Pump Applications. *IEEE Trans. Ind. Appl.* **2011**, *47*, 314–321. [CrossRef]
26. Lu, K.Y.; Rasmussen, P.O.; Watkins, S.J.; Blaabjerg, F. A New Low-Cost Hybrid Switched Reluctance Motor for Adjustable-Speed Pump Applications. In Proceedings of the 2006 IEEE Industry Applications Conference Forty-First IAS Annual Meeting, Tampa, FL, USA, 8–12 October 2006. [CrossRef]
27. Nakamura, K.; Ichinokura, O. Super-MultipolarPermanent Magnet Reluctance Generator Designed for Small-Scale Wind-Turbine Generation. *IEEE Trans. Magn.* **2012**, *48*, 3311–3314. [CrossRef]
28. Nakamura, K.; Ichinokura, O. Super-Multipolar Permanent Magnet Reluctance Generator Designed for Small-Scale Renewable Energy Generation. In Proceedings of the International Conference on Electrical Machines, ICEM, Marseille, France, 2–5 September 2012. [CrossRef]
29. Lee, C.; Krishnan, R. New Designs of a Two-Phase E-Core Switched Reluctance Machine by Optimizing the Magnetic Structure for a Specific Application: Concept, Design, and Analysis. *IEEE Trans. Ind. Appl.* **2009**, *45*, 1804–1814. [CrossRef]
30. Abhijith, V.; Hossain, M.J.; Lei, G.; A S, P. Performance Improvement of Switched Reluctance Motor Using Hybrid Excitation Method Without Permanent Magnets. In Proceedings of the 2021 24th International Conference on Electrical Machines and Systems (ICEMS), Gyeongju, Korea, 31 October–3 November 2021.

MDPI
St. Alban-Anlage 66
4052 Basel
Switzerland
Tel. +41 61 683 77 34
Fax +41 61 302 89 18
www.mdpi.com

Energies Editorial Office
E-mail: energies@mdpi.com
www.mdpi.com/journal/energies

www.ingramcontent.com/pod-product-compliance
Lightning Source LLC
Chambersburg PA
CBHW061547210926
51850CB00009B/2027
* 9 7 8 3 0 3 6 5 7 6 5 6 5 *